Ana Pérez Grassi

Variable illumination and invariant features for detecting and classifying varnish defects

**Forschungsberichte aus der Industriellen Informationstechnik
Band 1**

Institut für Industrielle Informationstechnik
Karlsruher Institut für Technologie

Hrsg. Prof. Dr.-Ing. Fernando Puente León
      Prof. Dr.-Ing. habil. Klaus Dostert

Variable illumination and invariant features for detecting and classifying varnish defects

by
Ana Pérez Grassi

KIT Scientific Publishing

Dissertation, Karlsruher Institut für Technologie
Fakultät für Elektrotechnik und Informationstechnik, 2010

Impressum

Karlsruher Institut für Technologie (KIT)
KIT Scientific Publishing
Straße am Forum 2
D-76131 Karlsruhe
www.ksp.kit.edu

KIT – Universität des Landes Baden-Württemberg und nationales
Forschungszentrum in der Helmholtz-Gemeinschaft

KIT Scientific Publishing 2010
Print on Demand

ISSN 2190-6629
ISBN 978-3-86644-537-6

Variable illumination and invariant features for detecting and classifying varnish defects

Zur Erlangung des akademischen Grades eines
DOKTOR-INGENIEURS
von der Fakultät für
Elektrotechnik und Informationstechnik
der Universität Karlsruhe (TH)
genehmigte
DISSERTATION
von

Ing. Ana Pérez Grassi

geb. in: Mendoza, Argentinien
Tag der mündlichen Prüfung: 17.12.2009
Hauptreferent: Prof. Dr.-Ing. Fernando Puente León
Korreferent: Prof. Dr.-Ing. Jürgen Beyerer

Contents

Acknowledgements

First of all, I sincerely thank Prof. Puente León for his guidance and for being constantly available whenever I needed it. He has been a great supervisor and mentor to me and I know that I always can count on his wise advice. I am particularly grateful to Prof. Beyerer for accepting to be my second examiner and for his valuable contributions to improving this thesis. To all my colleagues and the staff at VMS and MST (TUM), I am deeply grateful for their help and for the pleasant and friendly working atmosphere. Because of them, I will always look back on the time in Munich with affection. I profoundly thank my parents, Beatriz and Edmundo, who ever trusted and supported me. I know that they both would be proud of this achievement. I also thank my aunt Kitty, my brother Victor, my sisters Marcela and Elisa and the rest of my family for their unconditional support. I thank very much my husband and colleague Alejandro for the interesting discussions and for helping me find motivation during all my work.

Munich, June 2010

Abstract

This thesis presents a method to automatically detect and classify varnish defects on wood surfaces. In wood industry, the quality control of the varnish coat is performed by specialized workers. This results in a wastage of time, money, energy and a lot of unnecessary pollution, which can be significantly reduced by placing an automated control system at the end of the varnish line. However, varnish defects have some characteristics that make their automated recognition very difficult. The most important problem is that they are only partially visible under certain illumination directions. Then, one image of the surface does not contain enough information for a recognition task. On the contrary, a reliable classification requires to inspect the surface under different illumination directions, which results in the generation of a series of images. The relevant information for the classification is distributed along this series and should be extracted by defining a proper set of features. These features should be, on the one hand, enough sensitive to be able to distinguish among a big set of defect classes. On the other hand, they should be also enough robust in order not to be affected by the wood texture or by small local variations in the shape of defects. Moreover, they should be invariant against 2D Euclidean motion. In this thesis, such features are achieved by a novel method that associates the knowledge about the used illumination and defect shape to collect the relevant information from the series of images. The necessary feature invariance is reached by integrating over the transformation group and constructing histograms. Further, the integration over the transformation group is extended to consider series of images. The obtained invariant features have two dimensions and are classified by a Support Vector Machine. In conclusion, the proposed method results in invariant features that exhibit high discriminability and good generalization characteristics at the same time (90% correct classification and negligible false alarm rates). This work allows solving not only the problem of the automated recognition of varnish defects on wood surfaces, but it can also be easily adapted for inspecting other surfaces as well.

Zusammenfassung

Diese Dissertation präsentiert eine Methode zur automatischen Detektion und Klassifikation von Lackdefekten auf Holzoberflächen. Die Qualitätskontrolle dieser Oberflächen wird in der Holzindustrie von spezialisierten Mitarbeitern durchgeführt. Dies führt zur Verschwendung von Zeit, Geld und Energie, was durch den Einsatz eines automatischen Kontrollsystems beim Lackierverfahren stark verringert werden kann. Die optischen Eigenschaften von Lackdefekten erschweren jedoch ihre automatische Erkennung. Dabei ist es besonders problematisch, dass Lackdefekte unter bestimmten Lichtrichtungen nur teilweise sichtbar sind. Dadurch enthält ein einziges Bild nicht genug Information, um die Defekte korrekt zu erkennen. Aus diesem Grund müssen lackierte Holzoberflächen unter verschiedenen Lichtrichtungen untersucht werden, sodass eine zuverlässige Klassifikation vorgenommen werden kann. Hieraus ergibt sich eine Bildserie, welche, verteilt in ihren Bildern, die zur Defekterkennung relevante Information beinhaltet. Zur Extraktion dieser Information ist es notwendig, geeignete Merkmale zu definieren. Einerseits müssen diese Merkmale über eine ausreichend gute Empfindlichkeit verfügen, damit viele Defektklassen unterschieden werden können. Andererseits müssen die Merkmale auch genug Robustheit aufweisen, um weder von der Holztextur noch von lokalen Erscheinungsänderungen der Defekte beeinflusst zu werden. Eine weitere Herausforderung entsteht dadurch, dass Merkmale invariant bezüglich der 2D-Euklidischen Bewegung sein sollen (Defekte können an unterschiedlichen Stellen und mit verschiedenen Rotationswinkeln auftreten). Ein neuartiger Ansatz zur Ermittlung invarianter Merkmale aus beleuchtungsabhängigen Bildserien wird vorgestellt, der die Kenntnisse über die verschiedenen Lichtrichtungen und die Defektform zusammenfügt. Weiterhin wird die notwendige Invarianz bei den genannten Merkmalen durch die Integration über die Transformationsgruppe und die Bildung von Fuzzy-Histogrammen erreicht, wobei die Integration über die Transformationsgruppe für den Fall von Bildserien erweitert werden musste. Die daraus resultierenden invarianten Merkmale werden dann durch eine Support Vector Machine klassifiziert und führen sowohl zu einer hohen Diskrimination als auch zu einer sehr guten Generalisierung (eine Klassifizierungsrate von 90% und eine unerhebliche Anzahl an falschen Detektionen sind das Resultat). Die vorgeschlagene Methode ermöglicht nicht nur die automatische Erkennung von Lackdefekten auf Holzoberflächen erfolgreich zu bewerkstelligen, sondern kann auch mit wenig Aufwand zur Inspektion anderer Oberflächentypen angepasst werden.

1 Introduction

The varnishing of surfaces is practiced since millennia, however, it is still not a defect-free process. In spite of the current modern finishing systems and sophisticated coating techniques, finish defects cannot be completely avoided [SS96].

In the production of wood and wood-derived products, such as furniture and parquet slabs, the finishing process plays a very important role. The primary functions of any wood finish (e.g., paint, varnish, stain, etc.) are to protect the wood, to achieve a desirable appearance, and to provide a cleanable surface [Wil99]. A finish has to protect wood against abrasion, indentation, moisture, and possible color changes due to light or atmospheric pollutants. With respect to the surface appearance, there are different criteria depending on its use and personal preferences: Some products preserve wood in its natural state as much as possible, while others change its color and texture [Hoa00]. Thus, the selection of a finish depends on the desired appearance, necessary degree of protection and used substrate.

Once the correct finish is chosen, its application consists in the concatenation of more than a dozen of steps [Wil99, VM05]. The big number of variables involved in the coating selection and application makes the finishing process complex and susceptible to errors. A satisfactory finishing of wood surfaces can only be achieved when all factors that affect the process are given full consideration. These factors include characteristics of the wood substrate, properties of the finishing material, application techniques and severity of exposure [Wil99]. Any error in these factors or any wrong combination of them can cause defects on the coating film. This derives in dozens of defect types, where each one can be associated with different causes [SS96, GS02]. Moreover, the finishing process is an important source of pollution because of the repeated use of many dangerous chemicals. However, if already defective pieces are processed, all this pollution is unnecessary and should be avoided.

In addition, it can be distinguished between indoor and outdoor wood products. This work focuses on the inspection of indoor products, especially those manufactured by the furniture industry. In general, a finish for indoor products does not have to provide much protection against moisture, but it has to comply with more demanding standards of appearance and cleanability. This means that, for indoor wood products, varnish defects have a big esthetic significance. A much wider range of finish types and finishing methods is possible for indoor uses because weathering is not an issue and because of the many applications of the final product—from wood floors to cutting boards. Moreover, even restricting the inspection to only one product group such as furniture, many finishing methods and substrates are still possible.

Nowadays, the detection and classification of finish defects on wood surfaces is performed by specialized workers through visual inspection. This manual inspection causes a big loss of time and material, it is subjective and delivers non-reproducible results. In this thesis, an innovative method is proposed to automatically detect and classify finish defects on wood surfaces. An automated detection and classification allows reducing both the associated economic loss and the unnecessary pollution due to varnish defects. Further, the proposed method is able to deal with a big variability in color, texture, and shine of the inspected surface. This results in a robust inspection that can be effectively integrated into industrial production lines.

1.1 Motivation

As mentioned, varnished wood surfaces are currently inspected by human operators. This kind of inspection is inevitably associated with subjective results and is extremely time-demanding. As the classification criteria strongly depend on the operator, the product quality may suffer fluctuations. This hinders a standard classification of coating defects and therefore the establishment of uniform quality criteria in the furniture industry.

The finishing process consists of the concatenation of different steps [VM05], which are repeated a certain number of times until the desired final appearance is achieved. Between coating applications, the furniture pieces may be sanded, rubbed or polished and may be passed through drying ovens or flashoff areas. Normally, during this process, human operators perform only sporadic inspections. If defects are detected, it is probable that many repetitions of the finishing process have already been performed in vain. This yields a big loss of materials, time, work and energy. Additionally, as finishing is a continuous process that involves many furniture pieces at the same time, the later the defects are discovered, the more pieces are affected. Moreover, a late defect recognition can also hinder a quick determination of the cause, especially when this was sporadic. In this case, only a thorough and expensive laboratory analysis of the affected pieces can help identify the defects' cause in order to avoid it in the future.

Normally, defective pieces are reprocessed or discarded to be used in other applications. To reuse these pieces, the defective coating must first be removed. Such a coating removal is a complex procedure that depends on the used finish and wood type. Moreover, removing the coating from wood is only one half of the task, getting a paintable surface again is the other half [Wil99]. In addition to the loss of money, time and work with which this reprocessing is associated, the handling and disposal of the old coating can be restricted by severe regulations, particularly when it contains hazardous wastes like lead [Wil99]. On the other hand, not only removing the coating represents an ecological problem, but also the whole finishing process is extremely polluting. Finishing materials, such as stains and lacquers, are typically applied with help of a carrier solvent. The carrier solvent allows adjusting the finish to the right consistency and helps in the curing stage. In general, solvents are toxic to public health and to the environment [DKH$^+$92]. Depending on the type of coating, 70 to 90 percent of the liquid applied to a piece of furniture ends up as air emissions as the finish dries and/or cures. These emissions are rich in volatile organic compounds (VOCs) and hazardous air pollutant (HAP) [DKH$^+$92]. There are many sources of VOC/HAP emissions in the wood furniture industry: finishing, cleaning, mixing, gluing and touch-up and repair operations, but the finishing process normally accounts for the largest one [MF00]. Typical emitted pollutants include alcohols, methyl ethyl ketone (MEK), methyl isobutyl ketone (MIBK), toluene and xylene. Acetone, although not considered a VOC nor a HAP, is also emitted in large quantities [MF00]. Additionally, the solvent used to clean the equipment and to remove defective coatings in order to correct damaged pieces also generates emissions of VOC/HAP. Figure 1.1 describes all emission and waste sources from finishing and cleaning operations [MF00].

As previously mentioned, the finishing process involves a number of repetitions, which depends on the programmed number of coating applications (see Fig. 1.1). A quality control system at the end of each iteration can prevent defective pieces from receiving further coating applications and all the associated pollution. Additionally, if defective pieces are going to be reprocessed, the defective coating that should be removed will be thinner and in consequence less waste will be produced. In addition, the finishing process is also associated with large energy consumption (especially for the drying step where heat must be produced). The misuse of energy, when already defective pieces receive new coating films, leads not only to ecological damage, but also to an economic loss. In conclusion, an automated detection and classification of defects at the end of the finishing line will save up material, energy, time and work. Furthermore, it will reduce toxic emissions and wastes produced by unnecessary coating applications. In addition, defects will be detected and classified in an objective and reproducible way maintaining the product quality always at the desired level. The immediate detection and classification will allow stopping the production line before new pieces are affected. Thus, it will be possible to determine the exact scenario for which defects appear in order to isolate the cause. The fact that defects will not only be detected but also classified reduces the time invested in finding the problem. Identifying which classes of defects occur makes it possible to concentrate on searching solutions for a limited number of causes.

Currently, the lack of such an automated quality control system is due to the fact that recognizing coating defects on wood surfaces is a very complex task. Varnish defects are very small in comparison to the surface to be inspected and are only partially visible under certain illumination directions. Furthermore, the furniture industry uses a wide range of finishes and substrates. So, a quality control system should be able to recognize defects on surfaces with different appearances. This is especially critical when the coating is transparent and the wood texture is visible, since wood texture presents a big variability even inside the same wood species. For the recognition system, the visible wood texture constitutes a noise background that can mask defects and cause false alarms. Thus, the inspection results must be independent of the substrate texture. All these characteristics make a challenge out of the detection and even more the classification of defects on varnished wood surfaces, which is very difficult to solve with conventional pattern recognition and image processing methods.

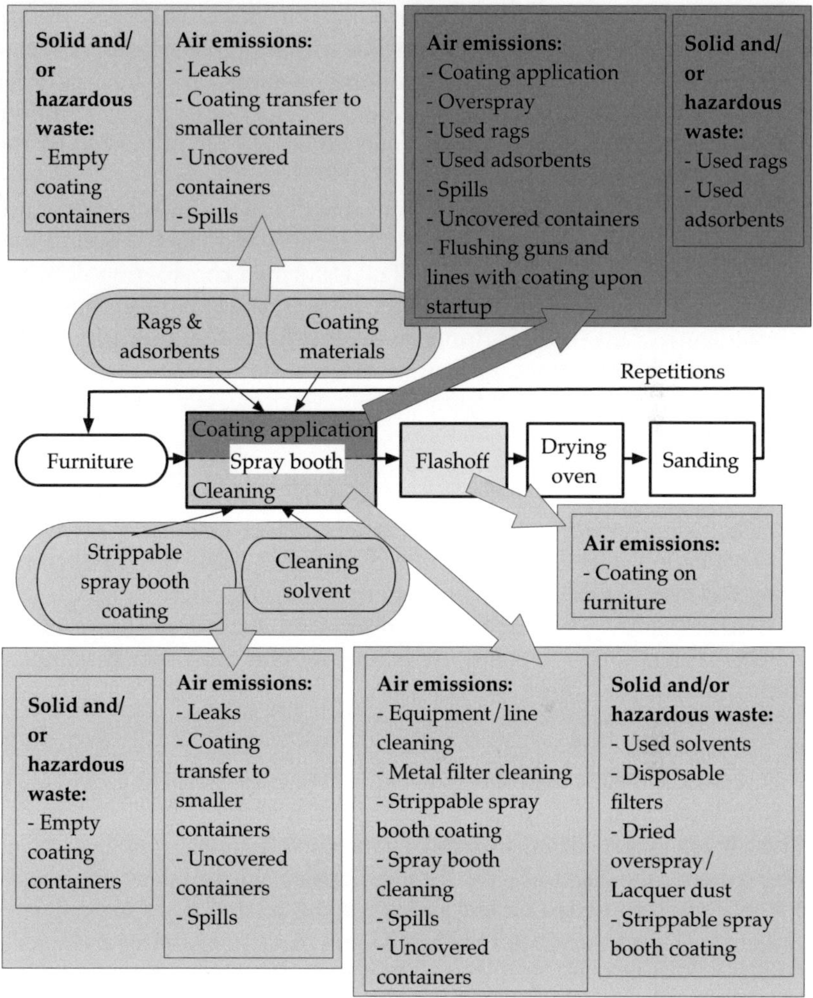

Figure 1.1: Emission and waste sources from finishing and cleaning [MF00].

1.2 Scope of this work

The objective of this thesis is to develop a method for detecting and classifying varnish defects on wood surfaces. The resulting method should present a high discriminability to be able to distinguish among different defect classes. However, at the same time, it should be robust in order not to be affected by small deviations on defect shapes. In addition, the detection and classification results should not be influenced by the texture of the substrate.

The proposed method combines variable illumination techniques and invariant features to achieve a good detection and classification ratio. While the image acquisition process is performed considering different illumination directions to collect all the relevant surface information, the construction of invariant features involves different techniques in order to achieve robustness and discriminability. Additionally, the invariant features are classified by a Support Vector Machine, which improves the ability of generalization of the whole method.

The main contribution of this thesis is the development of a method to extract invariant features from series of images taken under variable illumination directions. The construction of these features involves not only collecting relevant information distributed along the series of images, but also considering the influence of the illumination on the surface appearance. The features are optimized for recognizing varnish defects on wood surfaces, but the whole method can easily be adapted to the inspection of other materials.

1.3 Thesis organization

This thesis is structured in eight chapters and four appendices. The present chapter discusses the implications of finish defects for industry and environment. Chapter 2 presents the related work making special emphasis on those approaches that are more relevant for this thesis. Chapter 3 introduces briefly some concepts about pattern recognition. The acquisition process under variable illumination directions and the generation of series of images are described in Chapter 4. Chapter 5 presents the proposed method to construct invariant features from the obtained series of images. The classification process of the developed invariants is described in Chapter 6, while Chapter 7 presents and evaluates the results. Concluding remarks are further given in Chapter 8.

Appendix A presents some examples of series of images for different varnish defects. An introduction to the Support Vector Machine is further given in Appendix B, whereas Appendix C contains some images of the classification results. Finally, Appendix D gives a list of the used nomenclature.

2 Related work

In industry, surfaces of different materials, like wood, steel and stone, have to be inspected in order to detect defects. In the past two decades, there has been an increasing demand for automatizing such inspection processes. This has stimulated not only the research in the area of computer vision, but also the development of sensors, processors and illumination technologies. This chapter gives a brief overview of the related work in automated visual inspection with emphasis on methods and their industrial applications. Four relevant related research topics are described in separated sections: defect detection methods, invariant features by integration, classification methods and variable illumination techniques. Figure 2.1 illustrates the connection between the method proposed in this thesis and the related literature. The presented method to detect and classify finish defects is divided into three successive stages: image acquisition (Chapter 4), extraction of invariants (Chapter 5) and feature classification (Chapter 6). In Fig. 2.1, the arrows show the direct influence of approaches and techniques from the literature on the different stages of the proposed method. These approaches and techniques receive special attention in the remainder of this chapter.

2.1 Defect detection methods

This section discusses techniques for detecting defects on different types of surfaces. The following description is restricted to approaches based on texture analysis, where defects can be defined as local textural irregularities. Texture analysis methods can be grouped into four categories: statistical, structural, filter and model-based approaches.

2.1.1 Statistical approaches

Statistical approaches perform an indirect description of textures by means of the non-deterministic properties that characterize the distri-

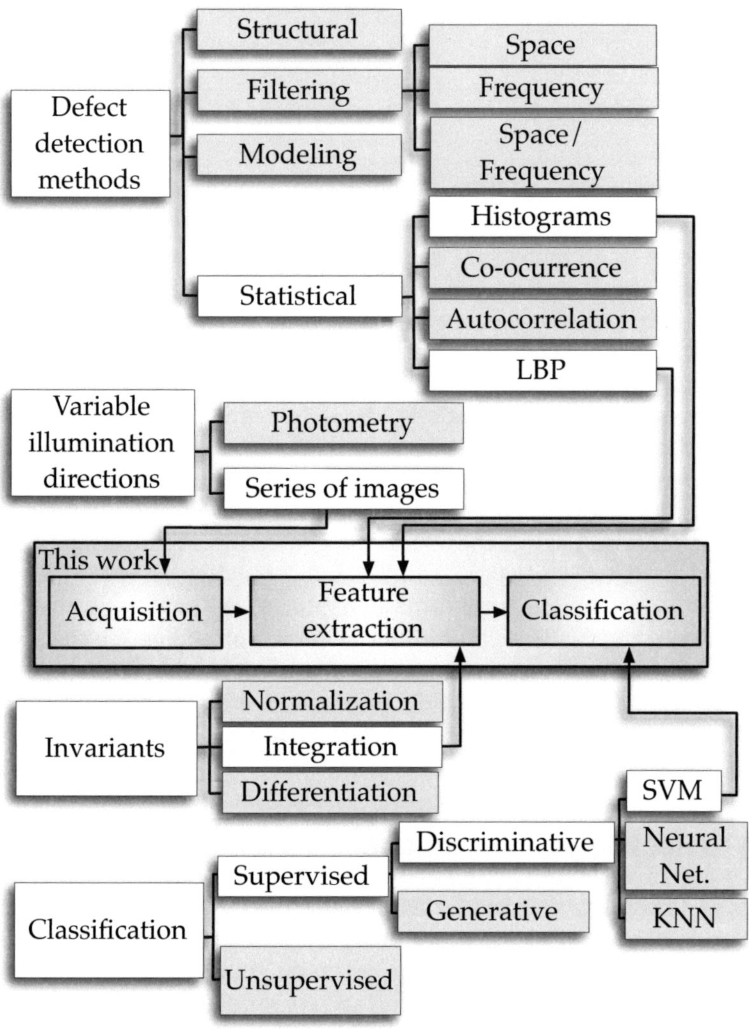

Figure 2.1: Related work areas and their influence on this work.

butions and relationships between the gray levels of an image. Further, first-order statistics involve a pixel-wise analysis, while higher order statistics require the study of pixel neighborhoods. These approaches present good results in inspecting inhomogeneous random textures (e.g. micro-textures), but they have problems in handling more structured textures, as they typically appear at a macroscopic level. This section concentrates on the statistical approaches that are most used in automated visual inspection: histograms, co-occurrence matrices, local binary patterns (LBP) and autocorrelation. As shown in Fig. 2.1, local binary pattern operators and histograms are particularly important for this thesis.

Histograms

For many applications, histograms have shown to be a useful and fast approach. The calculation of gray level histograms considers information regarding the single pixels rather than the relations between them. Thus, histograms contain the first order statistical information of images. Histograms present an intrinsic invariance against translation and rotation and are insensitive to the exact spatial distribution of gray levels. Common features based on histograms include mean, standard deviation, variance and median among others. Further, different histogram comparison methods can also be used to obtain features, for example, L_1 norm, L_2 norm, Mallows distance, divergence, histogram intersection, etc. Histograms are applied in several inspection processes. For example, color histograms have been used to detect shade defects in the production of tiles [CP00, XMT06], while local histograms have been applied to segment CT images in the area of medicine [BSPC05].

Co-occurrence matrices

Nowadays, co-occurrence matrices are one of the most well-known and widely used texture analysis approaches. They were introduced by Haralick et al., who illustrated their application to the classification of sandstone and the segmentation of satellite and aerial images [HSD73]. Co-occurrence matrices are based on second-order statistics, which capture the spatial dependency of two gray levels according to a displacement vector [HSD73]. Further, several textural features can be derived from co-occurrence matrices, such as energy, entropy, contrast, homogeneity and correlation. Co-occurrence matrices are used in a wide range of applications. Some examples are the detection of defects in paper webs [IRV96] and in woven textiles [BWBK97], and the classification of wool carpets according to their wear degree [SHW88]. More recently, Lindner et al. presented an extension of co-occurrence matrices to handle series of images, which was used to segment technical surfaces [LSP07].

Local binary pattern

Local binary patterns (LBP) are a very popular approach to texture analysis. It consists in describing a local texture through a binary code, which is obtained by thresholding a neighborhood by the gray level of its center pixel [OPH96]. This code can be regarded either as just a statistics of a non-linearly filtered image or as a micro-textural primitive. As a consequence, LBP can be considered to be a "unifying approach" between statistical and structural methods [MP05].

A rotation-invariant version of LBP, known as LBPROT, can be achieved by identifying all binary codes that match for some arbitrary cyclic shift [Oja96]. LBPROT was experimentally evaluated by Pietikäinen et al. in the classification of textures [POX00]. In order to improve LBP's ability to discriminate textures, Ojala et al. proposed to replace the thresholding by signed subtractions [OVOP01]. Furthermore, to achieve scale invariance, LBP was extended to a multiresolution approach, which consists in combining multiple LBP operators with different spatial resolutions and angular space quantizations [OPM02].

LBP's performance has been compared against other approaches such as co-occurrence matrices, Markov random field models, histograms and energy filters [OVOP01, GBC01, Iiv00]. In all cases, the most significant advantage of the LBP operator over other approaches is its better computational cost to performance ratio. An even faster algorithm to implement LBP was presented by Mäenpää [MTP03]. Among many other different applications, LBP has been used to inspect surfaces of different materials such as ceramics [Mon04], wood [NKS02, NSK01] and paper [MTP03]. A detailed analysis of the LBP operator together with possible extensions and applications can be found in [Mäe03].

Autocorrelation

Autocorrelation is an additional approach to texture analysis, which is motivated on the observation that many textures are regular by nature. The autocorrelation function of an image can be used to assess the amount of regularity as well as the fineness/coarseness of the texture in the image. If the texture is coarse, then the autocorrelation function will drop off slowly; otherwise, it will drop off very rapidly. In the case that the texture primitives are spatially periodic, the autocorrelation function will drop off and rise again in a periodic manner [LWY97, Har79]. Some examples of the use of autocorrelation can be found in the inspection of textiles [Woo90, HC04].

2.1.2 Structural approaches

Structural approaches characterize a texture by means of texture primitives and the spatial arrangement between them [VNP86]. The spatial relations between primitives can be obtained by modeling the geometry or by learning its statistical properties. The main drawback of structural approaches is that they are unable to capture the randomness of many textures. The most common structural approaches in defect detection rely on the analysis of primitive measures, edge extraction, skeleton representation and morphological operations. In the literature, there are several examples of structural approaches in industrial applications. For instance, texture primitive measures have been applied to detect defects in marble and granite [KMM$^+$94], edge extraction has been used in the inspection of leather [WX99] and skeleton representation has been employed in the quality control of textiles [CJ88]. Examples of morphological operations in detecting defects can be found in the inspection of textiles [MGD00] and of ceramics [EHE05]. An introduction to surface characterization by morphological filtering can be found in [PB05].

2.1.3 Filter-based approaches

Basically, filter-based approaches consist in applying filter banks to an image and then analyzing the filter responses, e.g. by computing their energy. These methods can be divided into spatial domain, frequency domain and joint spatial-frequency domain techniques.

Spatial domain filtering

The early attempts to describe textures by spatial domain filtering focussed on gradient calculations for extracting edges. Filters like Sobel, Robert's cross, Canny, Laplacian, Deriche, and Laws are well-established approaches to edge extraction. Another alternative for spatial domain texture analysis are the so-called eigenfilters. Some examples of spatial filters in detecting defects for industrial applications can be found in the inspection of textiles [KP02b] and ceramics [MMT04].

Frequency domain filtering

Many defect detection problems can be easier solved in the frequency domain. In this case, the image is transformed into the frequency domain using Fourier transform. The transformed image is then multiplied by the filter function in the frequency domain and, if necessary, transformed back to the spatial domain. In particular, frequency domain filters are used in the inspection of directional textures [TH99, Pue02a, BKP01, PR03], but also in the quality control of isotropic surfaces like sandpaper and leather [TH03].

Joint spatial-frequency domain filtering

Since Fourier coefficients depend on the entire image, the Fourier transform is not able to locate defective regions in the spatial domain. The classical way of introducing spatial dependency into Fourier analysis is through the windowed Fourier transform. If the window function is a Gaussian, the windowed Fourier transform becomes the well-known Gabor transform, which has been extensively studied in visual inspection. For industrial applications, the Gabor transform has been used, for example, in the inspection of steel [WPL00], textiles [KP00, BBL02, KP02a], sandpaper [TW00], stone and wood [KP02a]. Having similar properties as the Gabor transform, the wavelet transform is also widely used in defect detection. Some applications of wavelets can be found in the quality control of textiles [Ral02, YPY05, GSCS08], wood [TBH01], metal [MNA$^+$04] and stone [LM06]. An overview for industrial applications of the wavelet transform is presented by Truchetet and Laligant [TL08].

2.1.4 Model-based approaches

Model-based methods for texture analysis are concerned with the construction of an image model that can be used not only to describe a texture, but also to synthesize it. The model parameters must capture the essential properties of the texture [TJ99]. Model-based methods for defect detection include, among many others, fractal and random field models. Fractals are self-similar geometric primitives, which means that fragments of a fractal are exact or statistical copies of the whole fractal. Further, they can match it by being stretched and shifted. Many natural surfaces are self-similar at different scales. Such surfaces can be well modeled by means of fractals. Some examples of the use of fractals in detecting defects can be found in the inspection of textiles, leather [CP02] and painted surfaces [OD92]. Although fractals are suitable to model self-similar natural textures, they achieve a limited success in real applications. Markov random fields (MRFs) are also very popular in image modeling. They are able to capture the spatial contextual information in an image by assuming that the intensity at each pixel depends only on the intensities of the neighboring pixels. The MRF model performs well on micro-textures, but fails in describing regular and inhomogeneous textures. Industrial application of MRF can be found, for example, in the inspection of textiles [CFA91].

2.2 Classification

The primary goal of a visual inspection system is the detection and/or classification of defects. For example, in this thesis, the objective is to decide correctly whether a varnished wood surface presents a defect of a predefined class or not. Any classification approach involves choosing an appropriate decision scheme. According to the information needed by the classifier to construct this decision scheme, the classification approaches can be grouped into two categories: supervised and unsupervised. Different supervised and unsupervised classification techniques are described by Theodoridis and Koutroumbas [TK06].

2.2.1 Supervised classification

In supervised classification approaches the user has to classify manually a group of samples, which are then used to train the classifier. There are two basic methods to design a supervised classifier. The first one is based on so-called generative models. Generative classifiers learn a model of the joint probability of samples and classes. Then, the prediction for a new sample is made by applying Bayes rules to calculate the posterior probability and selecting the most likely class. The group of generative classifiers includes, among others, Bayesian networks and Gaussian classifiers. Industrial applications for generative classifiers can be found, for example, in the inspection of raw steel blocks using Bayesian networks [Per03].

The second method to design supervised classifiers is referred to as discriminative. Discriminative classifiers model the posterior probability, or learn a direct map from samples to classes. Examples of classifiers in this group include K-Nearest Neighbor (K-NN) classifiers, neural networks, perceptron, and Support Vector Machine (SVM). Some applications of K-NN classifier in industrial quality control include the inspection of ceramic [LVP05] and metal [MNA$^+$04] surfaces. Neural networks are extensively used due to their ability to learn complex non-linear input-output relationships. However, neural networks can show a poor ability to generalize for some applications. This drawback is overcome by the Support Vector Machine classifier. Some examples of industrial applications using neural networks can be found in the inspection of textiles [Kum03] and ceramics [MMT04]. SVM applications include, among others, the control of raw leather [PPMC06], steel [JMSC04] and textile [MBR04, Kar05]. Due to its better generalization ability, an SVM is applied in this thesis to classify invariant features. An overview of the SVM theory is presented in appendix B.

2.2.2 Unsupervised classification

In unsupervised classification approaches, the classifiers are trained using only the samples. The samples' classes are to be found by the classifier with no prior knowledge. This process, where the classifier groups samples into classes, is often called clustering. However, an unsupervised classification algorithm still requires the interaction with the user, but this time after the clustering has been performed. In this case, the user must assign to each created class its corresponding description. Examples of unsupervised classification methods include vector quantization and self-organizing maps.

2.3 Invariant features by integration

This work proposes a method to detect and classify defects on varnished wood surfaces based on the extraction of invariant features. There exist three fundamental methods to obtain invariant features: integration, differentiation and normalization [BS01]. The method presented in this thesis is based on the construction of invariants by integrating over the transformation group. These invariant features are known as integral invariants. The differentiation and normalization methods are out of the scope of this work (for more information, see [SM95a, BS01]).

The idea of obtaining invariance by integrating over the transformation group was born in the field of mathematics and group theory. The method was introduced by Hurwitz [Hur97] and rigorously proofed for locally compact Hausdorff topological groups by Haar [Haa33]. This latter work gives the integral over the transformation group the name of Haar integral. Von Neumann proved that the Haar measure is unique up to a multiplicative constant [vN36] and Weil [Wei40] and Cartan [Car40] extended the work for all locally compact groups. While the theory of integral invariants has been known for a long time, its application to pattern recognition is relatively new and receives currently quite a lot of attention. The application of group theoretical methods and invariance theory to image processing has been covered by Lenz [Len90]. Further, Schulz-Mirbach presented a detailed work about integral invariants in pattern recognition [SM95a]. The same author analyzed the effect of non-global transformations on integral invariants for classifying flexible objects [SM95b]. Additionally, the integration method can be generalized to consider subsets of transformation groups by changing the integration limits [HHB04]. More recently, Haasdonk et al. analyzed the adaptation of the Haar integral concept to obtain invariance in kernel machines like SVM [HVB05].

Integral invariants are extensively used in image retrieval, shape matching, defect detection and object classification. In the area of image retrieval, Siggelkow and Burkhardt successfully combined integral invariants with color histograms [SB98]. Using a Monte Carlo sampling method, the searching time in the previous method was significantly improved [SS99]. A more complete description of Siggelkow's work can be found in [Sig02]. Later, the Monte Carlo sampling was replaced with a salient points algorithm achieving an even better performance [HB04]. Additionally, Setia et al. extended this work with a relevance feedback method based on SVM classifiers [SIB05]. These image retrieval works have been used as basis for vision-based mobile robot localization techniques. Zhao et al. combined the integral invariants with particle filters to improve the localization capability of a mobile robot [ZKL08]. Furthermore, in the developing of global and local localization techniques, the Haar integral has been adapted to handle omnidirectional images [CLIM05, LICM06]. For shape matching, methods based on integral invariants have been developed for recognizing objects, which can be represented by closed planar contours [MCH$^+$06, LBH05]. These works were used to measure breast tumors on sonography images [YZC$^+$08], to classify fish species [LBH05] and even more, they were adapted to 3D data for face recognition [FKK07]. For defect detection, Schael presented an approach based on LBP and integral invariants [Sch05]. Schael modified the LBP operator replacing the thresholding by relational operations and used it as kernel function for the Haar integral [Sch01]. By combining the integral invariants with histograms, this method was successfully applied to the detection of textile defects, but it failed in their classification. For object classification, the Haar integral has been used, for example, in the recognition of handwriting [HHB04] and traffic participants [PTP$^+$07, TPFP08]. Furthermore, Schael and Siggelkow extended the Haar integral to handle 3D data [SS00]. Using this extension and an SVM, Ronneberger et al. classified pollen grains [RBS02]. Based on this last work, voxel-wise gray scale invariants were introduced and applied to the recognition of chorioallantoic membranes [RFB05]. Further, these voxel-wise gray scale invariants together with relevance feedback were used to segment and classify cell nuclei from 3D data [FRKB05, FB06]. Reisert and Burkhardt further extended the work in [RBS02] using directional information and spherical harmonic expansion [RB06]. This latter extension was used to classify proteins [TRB07] and 3D cell nuclei [SSR$^+$06]. An introduction to integral invariants is given later in Section 5.1.

2.4 Variable illumination direction

A visual inspection system depends entirely on the image acquisition procedure. In particular, the illumination plays a key role when imaging 3D textures. In this case, diffuse illumination can produce a destructive superposition of light and shadow resulting in an irreversible loss of information. For this reason, directional illumination seems more appropriate to inspect 3D textures. However, this kind of textures presents a different appearance under different illumination directions. So, the surface information captured in an image can drastically change when the illumination direction varies. This influence of the illumination direction on the image information has been analyzed in several works [CRL94, Cha94, Cha95, HLM06, Lin09]. Chantler and McGunnigle used sinusoidal functions of the illumination azimuth angle to model the effect of the illuminant rotation on the texture features [CM00]. This model was later formalized by showing that changes in the illumination azimuth angle cause texture features to describe Lissajous's ellipses [CSPM02]. Recently, Barsky and Petrou generalized this model for rough surfaces with variable albedo [BP07].

The fact that a 3D texture feature changes with the illumination direction makes its classification more difficult. However, a variable illumination direction can also be used to improve the knowledge about the texture. A well-known technique based on varying the illumination direction is photometric stereo, which uses a series of images to reconstruct the surface topography and albedo [Woo80]. This series of images is generated by varying the direction of the incident illumination between successive images while keeping the viewpoint constant. The number of images necessary to recover topography and albedo depends on the surface reflectance characteristics (for Lambertian surfaces, for example, only 3 images are sufficient). Based on photometric stereo, many approaches have been presented to classify 3D textures under different illumination directions. McGunnigle and Chantler proposed to extract the surface normal using photometric stereo in order to predict the texture appearance under different illumination directions. This is then used to train a texture classifier [McG98, MC00]. Bardera also used photometric stereo to classify textures under a varying illumination direction and variable sensor distances [LB03]. Further, Chantler and Wu introduced the concept of surface-rotation invariant (this concept will be analyzed and applied in Section 5.2.1) and used it in developing a texture classification approach [CW00, WC03]. Applying Lissajous's ellipse model from [CSPM02], Penirschke et al. developed a classifier for textures imaged under unknown illumination directions [PCP02]. In this case, a set of reference images of a texture was used to construct the Lissajous ellipse, from which the illuminant azimuth angle for a new image of the same texture can be extracted. For smooth surfaces with uniform Lambertian reflectance and shallow relief and under a sufficiently flat illumination, the effect of directional illumination can be eliminated by a bank of filters [DC05]. This latter approach was further used to construct an illumination-invariant texture classifier based on only one image.

For recognizing defects on 3D textures, many approaches perform a systematic variation of the illumination direction. For this purpose, sometimes, it is neither necessary nor efficient to reconstruct the surface topography. In this case, the illumination direction is systematically varied with the aim of recording all relevant surface information, which is then fused to recognize irregularities. The difference with photometric stereo is that the surface topography is not explicitly reconstructed, but its information is extracted and evaluated by generating and fusing series of images. Heizmann and Beyerer analyzed the optimal selection of capture parameters for the generation of series of images under variable illumination direction [HB05]. Additionally, Puente León presented a method to detect defects and measure certain characteristics on technical surfaces by combining a varying illumination direction and a multiple level fusion approach [Pue02b]. Further, Beyerer and Puente León proposed a controlled active image optimization method based on the fusion of series of images [BP05]. Two fusion approaches were proposed, an iterative and a global (final) one. The image signal to noise ratio (SNR) and local contrast were used as feedback measurements to control the image recording system. An example of the use of variable illumination directions in criminalistics can be found in the inspection of fired bullets to highlight striation marks left by firearms [Pue06]. In [PAP08], varnish defects on wood surfaces are classified from series of images by modeling their geometry. These models consider the defect shape and symmetry together with the used illumination direction. This leads to a bank of templates, which are correlated with the series of images. Finally, the results of the correlations are fused to perform the detection and classification. This method achieves a precise detection and classification of circular and linear defects, but it is computationally intensive. Different methods for surface segmentation using varying illumination direction were proposed by Lindner [Lin09].

3 Introduction to pattern recognition

This chapter presents basic concepts of pattern recognition, which are relevant for the approach presented in this work. It further introduces the notation used later in this thesis.

A recognition task is concerned with the categorization of *objects* according to some common properties. The categories, in which the objects are grouped, are called *classes* and are denoted by c. The action of assigning a class c to an object is called *classification* [GW02]. Let $\mathcal{O} = \{o^1, o^2, \ldots, o^{|\mathcal{O}|}\}$ be a set of objects; its classification can be described by an operator C_o in the following manner:

$$C_o : \mathcal{O} \longrightarrow \mathcal{C}; \quad o \mapsto c = C_o\{o\}, \tag{3.1}$$

where $\mathcal{C} = \{c^1, c^2, \ldots, c^{|\mathcal{C}|}\}$ is the set of object classes. The number of classes is given by $|\mathcal{C}| \in \mathbb{N}$. A simple representation of the object classification is shown in Fig. 3.1. If two or more objects belong to the same class c, there exists an equivalence relation between them. In general, two objects o^1 and o^2 in $\mathcal{O}$ are considered *equivalent*, if the result of their classification is identical:

$$o^1 \equiv o^2 \quad \Leftrightarrow \quad C_o\{o^1\} = C_o\{o^2\}. \tag{3.2}$$

As a consequence, the set of objects $\mathcal{O}$ can be divided in disjunctive subsets called equivalent classes $\mathcal{E}_c = \{o \mid C_o\{o\} = c\}$.

It is assumed that objects $o \in \mathcal{O}$ can be described using a limited number of measurements performed in finite time. These measurements allow representing objects through values, which can then be manipulated arithmetically. The measurement process can be represented by an operator S as follows:

$$S : \mathcal{O} \longrightarrow \mathcal{G}; \quad o \mapsto g(x) = S\{o\}, \tag{3.3}$$

where $\mathcal{G}$ is a vector space called *pattern space*. Elements $g(x) \in \mathcal{G}$ are denominated *patterns* and are functions on some set $\mathcal{X}$, with $x \in \mathcal{X}$. If $S\{o\} = g(x)$, it is said that $g(x)$ represents o. For the sake of simplicity,

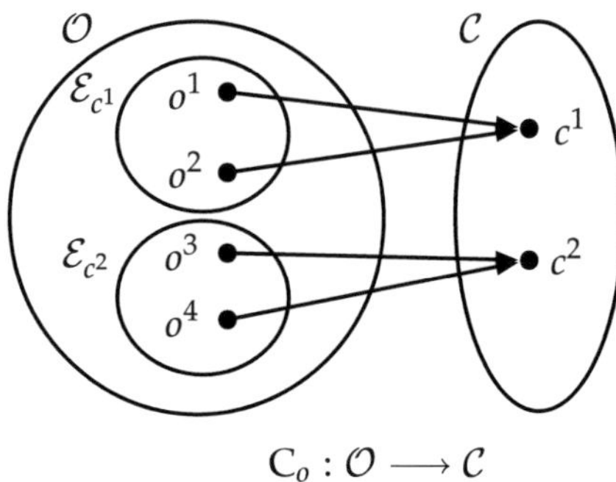

Figure 3.1: Classification of four objects into two classes ($|\mathcal{C}| = 2$).

the pattern argument is henceforth suppressed from the notation, i.e., $g := g(x)$. The pattern recognition task consists in identifying, from a pattern g, which equivalence class $\mathcal{E}_c$ the represented object o belongs to [Len90]. This can be described by a *pattern classification function* [Woo96]:

$$c_g : \mathcal{G} \longrightarrow \mathcal{C}; \quad g \mapsto c = c_g(g). \tag{3.4}$$

If neither noise nor other kinds of disturbances (e.g., defects of the sensor, changes of the sensor position, etc.) are considered in generating g, and if g contains all relevant information about the measured object, then it should be possible to define c_g (from Eqs. (3.1) and (3.4)) such that:

$$c_g(g) = C_o\{o\} \quad \Leftrightarrow \quad g = S\{o\}. \tag{3.5}$$

Figure 3.2 illustrates now the object recognition considering that patterns are ideally generated, i.e., without disturbances, and that the classification is performed in the pattern space $\mathcal{G}$. In practice, disturbances and loss of information may occur in the generation of patterns. In this case, Eq. (3.5) might not hold for all patterns, which results in false classifications. Consequently, generating patterns that are robust against disturbances is a major issue for a recognition task.

Although patterns represent objects, the direct classification of patterns is not always the most suitable way to recognize objects. In most cases, a pattern g contains much more information about the object o

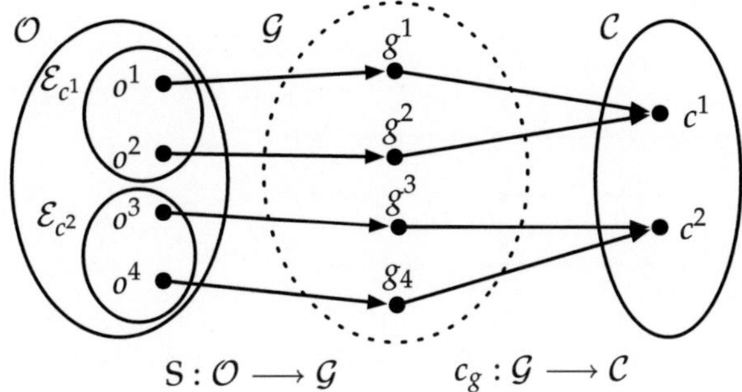

Figure 3.2: Generation and classification of four patterns. The objects are grouped into two equivalence classes: $\mathcal{E}_{c^1}$ and $\mathcal{E}_{c^2}$.

than what is relevant for the classification. Moreover, this excess of information can increase significantly the complexity of the recognition task. Thus, it is not always efficient, sometimes neither possible, to perform the classification directly in the pattern space. Hence, a function is defined to extract from each pattern only that information necessary for the classification [DHS01]:

$$f : \mathcal{G} \longrightarrow \mathcal{F}; \quad g \mapsto f(g). \tag{3.6}$$

$\mathcal{F}$ is a vector space called *feature space* and has in general smaller dimensions than $\mathcal{G}$. The functions $f(g) \in \mathcal{F}$ are denominated *features* and contain relevant information for the classification. The classification can now be performed in the feature space:

$$c_f : \mathcal{F} \longrightarrow \mathcal{C}; \quad f(g) \mapsto c. \tag{3.7}$$

Again, if disturbances can be neglected in the generation of patterns and features, and if features contain the relevant information for the classification, then it should be possible to define c_f (from Eqs. (3.5) and (3.7))such that:

$$c_f(f(g)) = c_g(g) = C_o\{o\} \quad \Leftrightarrow \quad g = S\{o\}.$$

Figure 3.3 illustrates the object recognition considering that patterns and features are ideally generated and extracted and that the classification is carried out in the feature space $\mathcal{F}$.

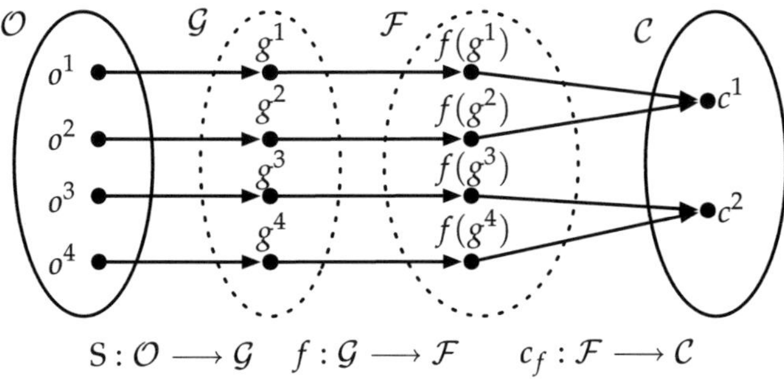

$$S: \mathcal{O} \longrightarrow \mathcal{G} \quad f: \mathcal{G} \longrightarrow \mathcal{F} \quad c_f: \mathcal{F} \longrightarrow \mathcal{C}$$

Figure 3.3: Feature extraction and classification.

Now, let the objects in the real world experiment certain transformations, like changes in their positions, sizes, etc. If these transformations do not alter the object properties that are relevant for the recognition task, then they should neither affect the classification results. For example, if the object position and size are not relevant for the classification, it does not matter which position or size an object has; it should always be assigned to the same class. In other words, all transformed versions of the object belong to the same equivalence class. However, through the measurement process, these transformations from the real world are induced in the pattern space $\mathcal{G}$ and should be considered in the classification. As a result, the classification process becomes even more complex. As described next, the classification can be simplified by extracting so-called invariant features, which are not affected by any induced transformation.

3.1 Classification by extracting invariants

As mentioned before, if an object undergoes a transformation in the real world, this induces a transformation in the pattern space $\mathcal{G}$. This latter is called *induced transformation*. The current section is concerned with obtaining features that are independent from induced transformations so that these do not affect the classification results. For obtaining such features, all objects of the same equivalence class must be related by the considered transformation. This way, $o^1 \in \mathcal{E}_c$ can be obtained by transforming any other object in the same equivalence class $\mathcal{E}_c$. Similarly, applying the induced transformation to any pattern of an object, it should be possible to generate all the patterns of the other objects in the same equivalence class. However, the induced transformation affects not only the object information in g, but also the disturbances introduced during the pattern generation. As a consequence, the resulting transformed pattern can only approximate the pattern of the transformed object. In what follows, this approximation is assumed to be *good enough* such that the transformed pattern and the pattern of the transformed object can be considered to be identical. As shown below, these characteristics give $\mathcal{G}$ the necessary structure to achieve features that do not change under the given induced transformation.

The *induced transformation* $\mathrm{T}(t)$ can be defined as a bijective map in $\mathcal{G}$ [Len90]:

$$\mathrm{T}: \mathcal{G} \times \mathcal{T} \longrightarrow \mathcal{G}; \quad (g, t) \mapsto \mathrm{T}(t)\{g\} \in \mathcal{G} \quad \forall g \in \mathcal{G}, t \in \mathcal{T}, \quad (3.8)$$

where $\mathcal{T}$ is the *set of transformation parameters* further denoted by t. The *set of all transformations* is given by $\mathrm{T}(\mathcal{T}) = \{\mathrm{T}(t) \,|\, t \in \mathcal{T}\}$. To simplify the notation, the braces of the transformation operator $\mathrm{T}(t)\{\}$ will be omitted: $\mathrm{T}(t)g := \mathrm{T}(t)\{g\}$.

If, for each transformation parameter $t \in \mathcal{T}$, there exists an inverse element $t^{-1} \in \mathcal{T}$ and $g = \mathrm{T}(t^{-1})\{\mathrm{T}(t)g\}$ holds, then the set of all transformations $\mathrm{T}(\mathcal{T})$ forms a group under the following composition rule [SM95a, Len90, Woo96]:

$$\mathrm{T}(t^1)\mathrm{T}(t^2)g = \mathrm{T}(t^1 \circ t^2)g \quad \forall t^1 t^2 \in \mathcal{T}, g \in \mathcal{G}. \quad (3.9)$$

In this case, $\mathrm{T}(\mathcal{T})$ is called *transformation group*. The symbol $\circ$ represents a mathematical operation. For example, if $\mathrm{T}(\mathcal{T})$ describes a two-dimensional rotation and t the corresponding angles, then $\circ$ represents the sum of these angles. If this group describes all induced transformations that are not relevant for the classification, all patterns $g \in \mathcal{G}$ that can be associated through transformations in $\mathrm{T}(\mathcal{T})$ describe objects in the same equivalence class $\mathcal{E}_c$. Moreover, they are related by an equivalence relation, which can defined as [Dun73, Len90]:

$$g^1 \stackrel{\mathrm{T}(\mathcal{T})}{=} g^2 \quad \Leftrightarrow \quad g^2 = \mathrm{T}(t)g^1, \quad \text{for some } t \in \mathcal{T}. \quad (3.10)$$

It is considered that all patterns associated by the given $\mathrm{T}(t)$ represent objects of the same equivalence class. As mentioned above, transformed patterns and patterns of transformed objects are considered identical. Therefore, the classification function yield the same result for both of them:

$$c_g(g) = c_g(T(t)g) \quad \forall t \in \mathcal{T}, g \in \mathcal{G}.$$

For simplifying the notation, the equivalence relation $g^1 \overset{T(\mathcal{T})}{=} g^2$ can be written as $g^1 \equiv g^2$. However, it must be taken into account that this equivalence is defined by the induced transformation group $T(\mathcal{T})$. This relation divides the space $\mathcal{G}$ in equivalence classes $\mathcal{E}_p$, where p is a pattern from $\mathcal{G}$ called *prototype*:

$$\mathcal{E}_p = \{g \in \mathcal{G} \,|\, g = T(t)p\} \quad \forall t \in \mathcal{T}. \tag{3.11}$$

In case that patterns are generated without any disturbances and contain all the relevant information for the classification, each equivalence class $\mathcal{E}_p$ in the pattern space coincides with an equivalence class $\mathcal{E}_c$ in the object space.

The set of all equivalence classes in $\mathcal{G}$ is denoted by $\mathcal{G}/T(\mathcal{T})$. Now, the space $\mathcal{G}$ can be defined as the disjoint union of the different equivalence classes $\cup_{\mathcal{G}/T(\mathcal{T})} \mathcal{E}_p = \mathcal{G}$. Figure 3.4 illustrates with an example the resulting structure of $\mathcal{G}$.

The space $\mathcal{G}$ composed by the union of disjoint equivalence classes $\mathcal{E}_p$ (see Eq. (3.11)) presents the structure described at the beginning of this section. This structure of the pattern space allows constructing features that are insensitive to the transformations $T(t) \in T(\mathcal{T})$. This kind of features denoted by $\tilde{f}(g)$ are called *invariants*. A feature is an invariant under a transformation group $T(\mathcal{T})$, if and only if, it remains constant on each equivalence class $\mathcal{E}_p$ [Dun73]:

$$\tilde{f}(T(t)g) = \tilde{f}(g) \quad \forall\, g \in \mathcal{G}, t \in \mathcal{T}.$$

A feature is denominated *relative invariant* with weight or modulus χ, if it is not absolutely invariant but it depends on the transformation $T(t)$ by a multiplicative factor:

$$f(T(t)g) = \chi(t)f(g).$$

If the weight χ is trivial, i.e., $\chi(t) = 1 \,\forall t \in \mathcal{T}$, the relative invariant feature becomes an invariant one [Woo96, DC70].

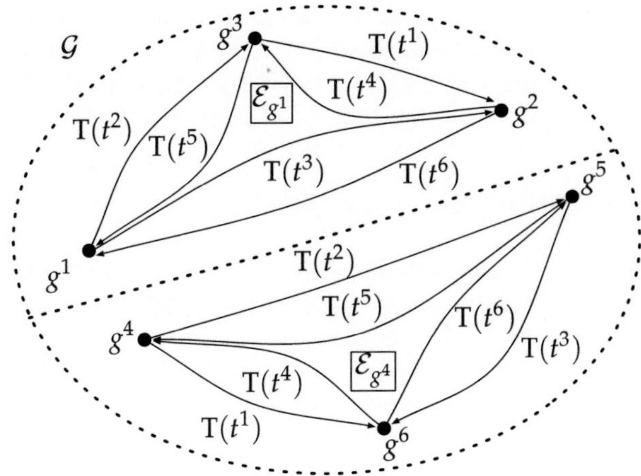

Figure 3.4: The pattern space $\mathcal{G}$ is partitioned into two equivalence classes ($|\mathcal{G}/\text{T}(\mathcal{T})| = |\mathcal{C}| = 2$), where g^1 and g^4 are the corresponding prototypes and $|\mathcal{T}| = 6$.

A feature $\tilde{f}(g)$ is a *complete invariant* under $\text{T}(\mathcal{T})$, if and only if, $\tilde{f}(g)$ is invariant and has different values on different equivalence classes $\mathcal{E}_p$:

$$\tilde{f}(g^1) = \tilde{f}(g^2) \quad \Leftrightarrow \quad g^1 \equiv g^2 \quad \forall\, g \in \mathcal{G},\, t \in \mathcal{T}.$$

In this sense, a complete invariant is able to separate the equivalence classes induced by $\text{T}(\mathcal{T})$ [Dun73]. Complete invariants are typically built upon a sequence of many individually incomplete invariants [SSY79, Woo96]. A sequence $\tilde{\mathcal{F}}(g) := \left\langle \tilde{f}^l(g), l \in \{1,\dots,L\} \right\rangle$ of invariants under $\text{T}(\mathcal{T})$ is called a *complete system of invariants*, if, for g^1 and $g^2 \in \mathcal{G}$, there exists an integer L such that the following holds for some $\varepsilon > 0$:

$$\tilde{f}^l(g^1) = \tilde{f}^l(g^2) \Rightarrow \exists\, t \in \mathcal{T} \mid \|g^1 - \text{T}(t)g^2\| < \varepsilon \quad \forall\, l. \tag{3.12}$$

Equation (3.12) indicates that a complete system of invariants may distinguish with arbitrary accuracy between any two non-equivalent patterns. In this case, it is said that the system presents *perfect discriminability*.

There are three fundamental methods to obtain invariant features: integration, differentiation and normalization [BS01]. The integration method, which is introduced in Section 5.1, is the one used in this thesis.

3.2 Application-oriented definitions

Most of the concepts presented in the previous sections acquire a more concrete meaning when an application is considered. In this thesis, the recognition task consists in detecting and classifying varnish defects on wood surfaces through a vision system. The proposed method is based on varying illumination direction and extracting of invariants features. This section presents a list of application-oriented definitions that are used along all this thesis.

- Objects $o \in \mathcal{O}$:

  The objects of interest are regions of varnished wood surfaces of a defined size.

- Classes $c \in \mathcal{C}$:

  The surface regions are classified according to the defects that they show (this includes the absence of defects). The classes considered in this thesis are: crater, fissure, bubble, pickle, pin-holing, peak and no defect ($|\mathcal{C}| = 7$). Descriptions and examples of these defects are presented later.

- Pattern $g \in \mathcal{G}$ and measurement operator S:

  The varnished surfaces are inspected with a camera. Hence, the patterns are images, which can be described as two dimensional functions $g(\mathbf{x})$, where $\mathbf{x} = (x, y)^{\mathrm{T}}$ and $\mathbf{x} \in \mathcal{X}^2$. The image $g(\mathbf{x})$ has a finite size given by $0 \leq x < M'$ and $0 \leq y < N'$. Moreover, the pattern $g(\mathbf{x})$ depends not only on the object o but also on the image acquisition conditions, which can be completely described by the sensor, optics and illumination parameters. If the vector $\mathbf{k}$ describes all acquisition parameters, then the measurement operator and the resulting pattern in Eq. (3.3) can be rewritten as:

$$\mathrm{S} : \mathcal{O} \times \mathcal{K} \longrightarrow \mathcal{G}; \ (o, \mathbf{k}) \mapsto g(\mathbf{x}, \mathbf{k}) = \mathrm{S}\{o, \mathbf{k}\}, \qquad (3.13)$$

where $\mathcal{K}$ is the set of all possible acquisition parameter vectors $\mathbf{k}$.

If the used camera is a digital one, then the pattern is not a continuous function on $\mathcal{X}^2$, but a discrete one. A discrete image is defined as follows:

$$g_{mn}(\mathbf{k}) := g(m\Delta x, n\Delta y, \mathbf{k}), \tag{3.14}$$

with $m \in \{0, \ldots, M-1\}$, $n \in \{0, \ldots, N-1\}$, $M = \frac{M'}{\Delta x}$ and $N = \frac{N'}{\Delta y}$.

When the acquisition conditions are not relevant for the analysis, the vector $\mathbf{k}$ is suppressed from the notation. In this case, patterns will be denoted by $g(\mathbf{x})$ in the continuous case and g_{mn} in the discrete one. From the parameters in $\mathbf{k}$, those related to the illumination direction are of special interest for this thesis and are analyzed in Chapter 4.

- Transformation group $\mathrm{T}(\mathcal{T})$:

  Finish defects can be affected by a rigid two-dimensional (2D) motion, which should not alter their classification. This motion is characterized by a translation (defects can appear at different locations on the surfaces) and a rotation. As a consequence, for this thesis, the transformation group $\mathrm{T}(\mathcal{T})$ of interest is the 2D Euclidean motion. For the sake of simplicity, the description of this transformation group begins considering only a 2D rotation. Thus, an induced 2D rotation in $\mathcal{G}$ can be described in the following manner:

  $$\mathrm{R} : \mathcal{G} \times \mathcal{A} \longrightarrow \mathcal{G}; \; g(\mathbf{x}) \mapsto \mathrm{R}(\varphi)g(\mathbf{x}),$$

  where $\mathcal{A}$ is the set of all rotation angles φ, with $\varphi \in [0, 2\pi)$. The transformation $\mathrm{R}(\varphi)$ can be described as follows:

  $$\mathrm{R}(\varphi)g(\mathbf{x}) = g\begin{pmatrix} x\cos\varphi - y\sin\varphi \\ x\sin\varphi + y\cos\varphi \end{pmatrix} \quad \forall \varphi \in \mathcal{A}. \tag{3.15}$$

  The 2D rotation constitutes a group under the usual composition rule (see Eq. (3.9)). This transformation group, denoted by $\mathrm{R}(\mathcal{A})^1$,

---

[1]In the literature, the group of the 2D rotation is usually denoted by $SO(2)$.

is compact, i.e., bounded and closed as explained later. (Compactness is a necessary condition to obtain invariants by integration.)

The 2D Euclidean motion transformation is constituted by a rotation followed by a translation. The 2D Euclidean motion on $\mathcal{G}$ can be defined by the following mapping:

$$\mathrm{M}: \mathcal{G} \times \mathcal{M} \longrightarrow \mathcal{G}; \quad g(\mathbf{x}) \mapsto \mathrm{M}(\tau_x, \tau_y, \varphi)g(\mathbf{x}).$$

$\mathcal{M}$ is the set of all transformation parameter vectors $(\tau_x, \tau_y, \varphi)$. The parameters τ_x and τ_y describe respectively the translation in direction x and y, while φ represents the rotation angle. The transformation $\mathrm{M}(\tau_x, \tau_y, \varphi)$ is defined as follows:

$$\mathrm{M}(\tau_x, \tau_y, \varphi)g(\mathbf{x}) = g \begin{pmatrix} x \cos \varphi - y \sin \varphi + \tau_x \\ x \sin \varphi + y \cos \varphi + \tau_y \end{pmatrix}. \qquad (3.16)$$

The set of all 2D Euclidean motion transformations is denoted by $\mathrm{M}(\mathcal{M})^2$. This set forms a group under the usual composition rule (Eq. (3.9)). However, $\mathrm{M}(\mathcal{M})$ is not a compact group [Vil68]. To be compact, a group must be bounded and closed. As the translation parameters τ_x and τ_y are defined in the interval $(-\infty, +\infty)$, this group is clearly not bounded. Fortunately, as the images have a finite size $M' \times N'$, the translation parameters can be limited to $\tau_x \in [0, M')$ and $\tau_y \in [0, N')$. This delimitation allows bounding the transformation group. On the other hand, considering a finite signal results in a loss of closeness. Figure 3.5 illustrates this problem with an image of a triangle-shaped object. Let the geometrical shapes of objects be the property used to recognize them and therefore to compose the equivalence classes. In this case, the object belongs to the class "triangle" (Figure 3.5(a)). Any translation (Fig. 3.5(b)) or rotation (Fig. 3.5(c)) for which a part of the object exceeds the image borders, results in the object being cut. This cut alters the original object shape and the resulting transformed pattern will not belong to the same equivalence class anymore. In this example, after the cut, the triangle will not be a triangle anymore.

---

[2]In the literature, this group is usually denoted by $M(2)$.

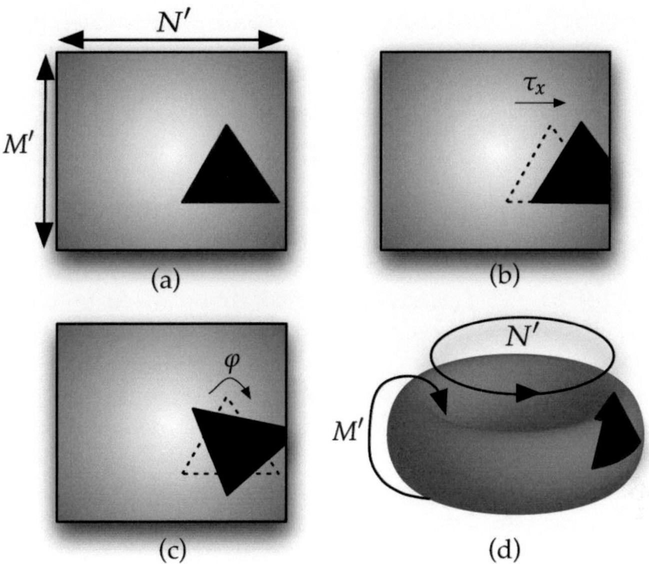

Figure 3.5: Representation of the cyclical 2D Euclidean transformation:
(a) untransformed object, (b) object cut by translation, (c) object cut by
rotation, (d) closeness by cyclical rotation and translation.

However, the closeness of the group (and therefore the structure
of Fig. 3.4) can be recovered by defining the translation and rotation as cyclical. This latter can be interpreted as the junction of the
image contours, which results in a toroid (Fig. 3.5(d)) rather than
a plane (Fig. 3.5(a)). Now, no transformation in M($\mathcal{M}$) changes
the equivalence class of the pattern. Under these conditions, the
cyclical 2D Euclidean transformation becomes a compact group,
which can be defined as follows:

$$M(\tau_x, \tau_y, \varphi)g(\mathbf{x}) = g\begin{pmatrix}(x\cos\varphi - y\sin\varphi + \tau_x)\bmod M' \\ (x\sin\varphi + y\cos\varphi + \tau_y)\bmod N'\end{pmatrix}. \quad (3.17)$$

If the pattern is a discrete function as defined in Eq. (3.14),
the transformation parameter vector is given by $(i\Delta x, j\Delta y, k\Delta\varphi)$,
where $i \in \{0,\ldots,M-1\}, j \in \{0,\ldots,N-1\}$ and $k \in \{0,\ldots,K-$

1} with $K = 2\pi/\Delta\varphi$. The discrete cyclical 2D Euclidean transformation is denoted by:

$$\mathrm{M}_{ijk}\, g_{mn} = g \begin{pmatrix} ((m \cos k_\varphi - n \sin k_\varphi + i) \bmod M)\Delta x \\ ((m \sin k_\varphi + n \cos k_\varphi + j) \bmod N)\Delta y \end{pmatrix}, \quad (3.18)$$

with $k_\varphi := k\Delta\varphi$, and the discrete cyclical 2D rotation transformation can be written as:

$$\mathrm{R}_k\, g_{mn} = g \begin{pmatrix} (m \cos k_\varphi - n \sin k_\varphi) \bmod M)\Delta x \\ (m \sin k_\varphi + n \cos k_\varphi) \bmod N)\Delta y \end{pmatrix}. \quad (3.19)$$

The pattern recognition concepts discussed previously are used in the following chapters to develop a method for obtaining invariant features. These invariants are shown to be able to represent the varnished surfaces with enough robustness and discriminability so as to perform a detection and classification of defects on their basis.

4 Image acquisition under variable illumination

As discussed previously, the patterns are functions of the imaged object and the used acquisition parameters $g(\mathbf{x}, \mathbf{k})$ (see Eq. (3.13)). Among the elements of $\mathbf{k}$, the illumination direction is of special interest for this thesis. Before analyzing the effect of the illumination on the images, the illumination conditions are described. From now on, it is assumed that the used light source delivers a collimated, distant and punctual illumination. This ensures a local constant light radiation on planar surfaces [Pue99]. For describing the illuminant direction, a spherical coordinate system is used to characterize the space around the inspected surface. As only opaque surfaces are considered, from this space only the upper hemisphere, whose equator matches the plane tangent to the surface, is relevant for the image acquisition analysis (see Fig. 4.1). Each point on this hemisphere describes a possible light source position. The coor-

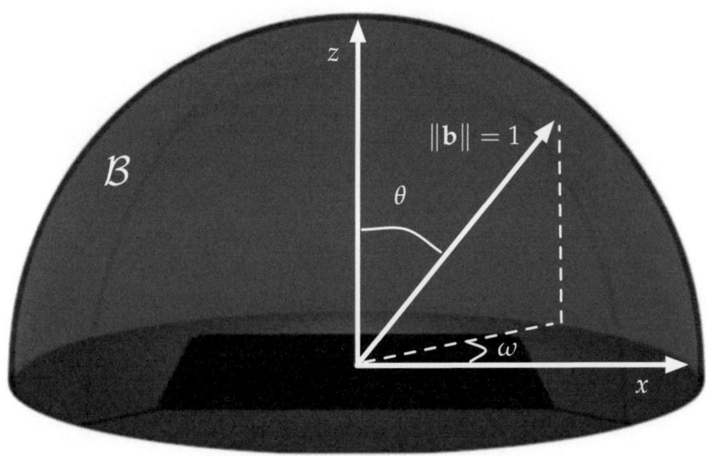

Figure 4.1: Illumination space $\mathcal{B}$.

dinates to locate these points are given by the corresponding azimuth angle $\omega \in [0, 2\pi)$, elevation angle $\theta \in [0, \pi/2]$ and radius $r \in [0, \infty)$. The assumption of a distant illumination source allows fixing $r = \infty$. Then, all illumination directions can be described through unit vectors $\mathbf{b} = (\omega, \theta)$. The space $\mathcal{B}$ composed by all unit vectors $\mathbf{b}$ is called *illumination space* [Lan91]. In conclusion, during the image acquisition, the inspected surface is illuminated from a direction $\mathbf{b}$ by means of a collimated distant punctual light source.

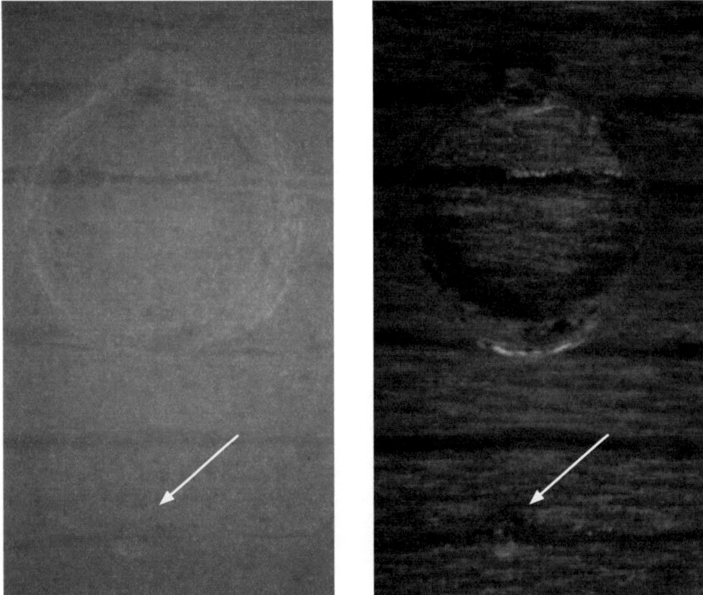

Figure 4.2: Craters under diffuse (left) and directional (right) light.

Now, the effect of the light on the surface and therefore on the image is going to be analyzed. First, the camera should be placed in the scene. Let the camera be fix with its optical axis parallel to the axis z (Fig. 4.1). Thus, the images captured by the camera show a top view of the surface. If the surface is lighted from a particular direction $\mathbf{b} \in \mathcal{B}$, the illumination is called *directional*. On the other hand, the illumination is called *diffuse*, if the surface is lighted at the same time from all directions

$\mathbf{b} \in \mathcal{B}$. Diffuse illumination can result in a destructive overlapping of light and shadow, which can cause a decrease of the image contrast and loss of information [Pue99]. To exemplify this effect, Fig. 4.2 shows two pictures of the same varnished surface under diffuse illumination (left) and under directional illumination (right). The varnish film showed in the figure is damaged and contains two craters. While under diffuse illumination both defects are practically imperceptible, under directional illumination they can be partially recognized. Under directional illumination, the local image contrast is higher; however, only some parts of the defects are clearly visible. Which parts of the defects become visible depends on the surface normal $\mathbf{n}$ and the illumination direction $\mathbf{b}$.

Under directional illumination, the varnish film behaves partially like a mirror. Then, the incident light beam is predominantly reflected in a direction depending on the local normal vector $\mathbf{n}$ of the surface. A non-defective surface can be assumed to be perfectly planar. In this case, the optical axis of the camera coincides with the normal vector $\mathbf{n}$ (Fig. 4.3(a)). If $\mathbf{n}$ and $\mathbf{b}$ are not parallel, a non-defective surface appears dark to the camera. On the other hand, in the presence of defects, the reflected light beam might be partially spread towards the camera due to the local variation of the normal vector $\mathbf{n}$ (Fig. 4.3(b)). This effect becomes visible in the form of highlights on the defect boundaries.

From all parameters in $\mathbf{k}$, only the illumination direction given by $\mathbf{b}$ is of interest in this thesis. All other elements of $\mathbf{k}$ are considered constant and suppressed from the notation. Now, the pattern is defined as an image taken under certain illumination direction given by $\mathbf{b}$:

$$S\{o, \mathbf{b}\} = g(\mathbf{x}, \mathbf{b}).$$

The objective of this thesis is to classify the surfaces according to the defects that they exhibit. As defects are three dimensional, their information is contained in the surface topography. However, as mentioned before, defects are partially visible under directional illumination. Then, the information about the surface topography contained in one image $g(\mathbf{x}, \mathbf{b})$ is incomplete and insufficient for a classification task. In order to obtain a suitable representation of the surface in the pattern space, it is necessary to collect all its relevant information. This latter requires the inspection of the surface under different illumination directions, which further implies a sampling of the illumination space $\mathcal{B}$.

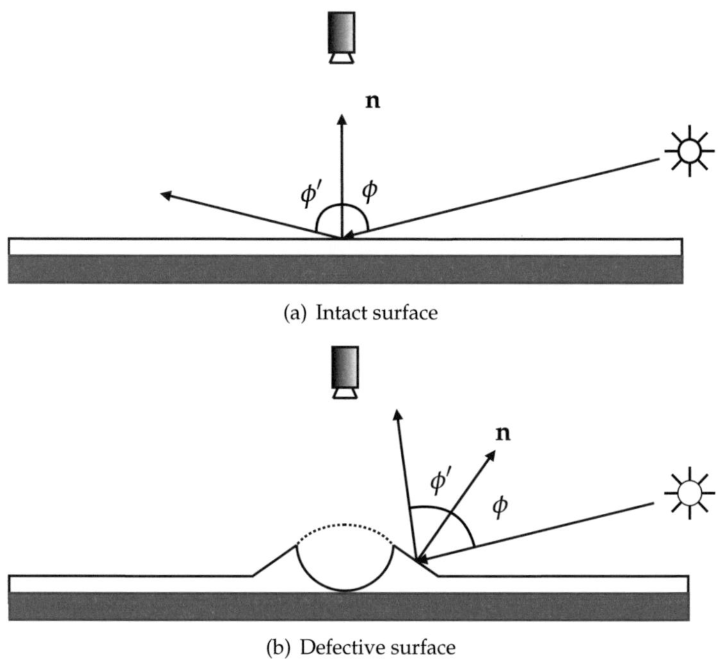

(a) Intact surface

(b) Defective surface

Figure 4.3: Reflection characteristics of an intact and a defective surface.

Taking the camera position and the possible variations of **n** into account, it is possible to limit the sampling of $\mathcal{B}$ to a subspace $\mathcal{B}_{\theta'} \subset \mathcal{B}$. This subspace $\mathcal{B}_{\theta'}$ is composed of all illumination directions $\mathbf{b} = (\omega, \theta')$ with a constant elevation angle θ' (see Fig. 4.4). The elevation angle can be fixed to θ', because for the selected camera position only a small range of angles ϕ between the incident light beam and **n** results in a deviation of the reflected beam towards the sensor (see Fig. 4.3(b)). As the elevation angle is considered constant, it can be suppressed from the notation and the pattern can be rewritten as $g(\mathbf{x}, \omega) := g(\mathbf{x}, \mathbf{b})$. The sequence of all images that results from a continuous acquisition, while successively varying the illumination direction along $\mathcal{B}_{\theta'}$, is called *series of images* and denoted by:

$$\mathcal{S}_g = \langle g(\mathbf{x}, \omega),\ 0 \le \omega < 2\pi \rangle. \tag{4.1}$$

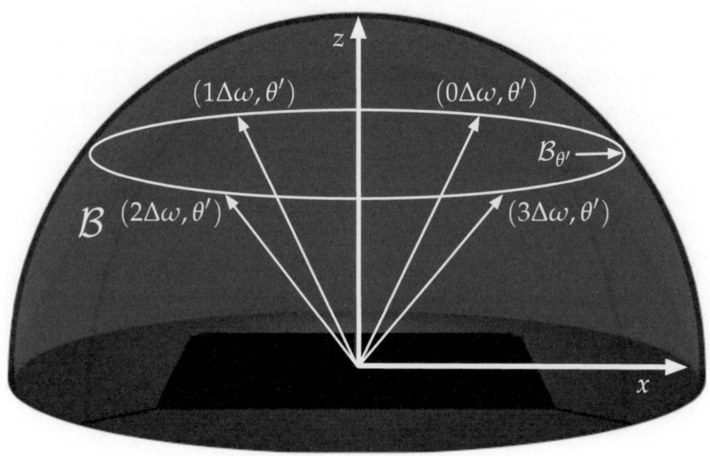

Figure 4.4: Sampling of the illumination space $\mathcal{B}$ with $\Delta\omega = \pi/2$ ($B = 4$).

In the practice, the image acquisition is not performed while varying the illumination azimuth in a continuous way, but by sampling the subspace $\mathcal{B}_{\theta'}$ with a frequency $1/\Delta\omega$. This frequency should be low enough to allow practical applications. At the same time, it should be high enough to collect all the relevant information of the surface. An analysis of how to select the value of $1/\Delta\omega$ can further be found in [LP06]. This sampling results in a subset of illumination directions denoted by $\mathcal{B}' \subset \mathcal{B}_{\theta'}$ and is defined as follows:

$$\mathcal{B}' = \{\mathbf{b} = (b\Delta\omega, \theta')|\, b \in \{0, \ldots, B-1\}\}, \tag{4.2}$$

with $B = 2\pi/\Delta\omega$ (Fig. 4.4). Then, for each illumination direction $\mathbf{b} = (b\Delta\omega, \theta')$ an image $g(\mathbf{x}, b\Delta\omega)$ is recorded. For a given $\Delta\omega$, the pattern can be more easily written as $g(\mathbf{x}, b) := g(\mathbf{x}, b\Delta\omega)$. In this case, the series of images is defined as:

$$\mathcal{S}_g = \langle g(\mathbf{x}, b),\, b \in \{0, \ldots, B-1\}\rangle.$$

For a digital sensor, the series of images is described as follows according to Eq. (3.14):

$$\mathcal{S}_g = \langle g_{mnb},\, b \in \{0, 1, \ldots, B-1\}\rangle. \tag{4.3}$$

Throughout this thesis, the denomination *continuous series of images* is going to be used for Eq. (4.1) and *discrete series of images* for Eq. (4.3). An example of a discrete series of images $\mathcal{S}_g$ is showed in Fig. 4.5. More examples of discrete series of images of the different defect classes are given in Appendix A.

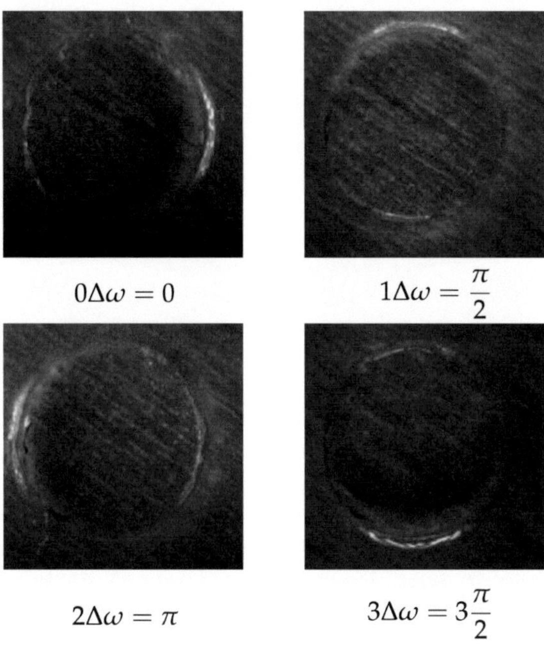

Figure 4.5: Series of images of a varnished wood surface showing a crater for $B = 4$. The acquisition parameters are $\Delta\omega = 90°$ and $\theta' \approx 80°$.

In practice, the sampling of the illumination space is achieved by rotating the illumination source around the surface an taking a picture after each displacement equal to $\Delta\omega$. Figure 4.6 illustrates the acquisition setup.

For a classification task, the series of images (and not each individual image) contains the whole relevant information about the surface. Therefore, the complete series of images represents the surface in the pattern space and constitutes the real pattern. This modifies the mea-

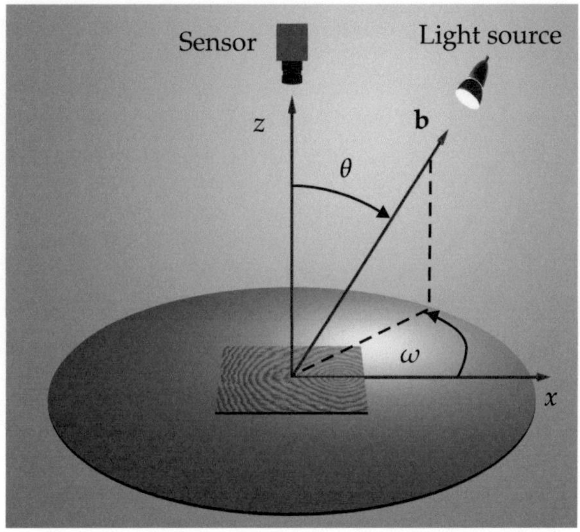

Figure 4.6: Image acquisition setup.

surement operator definition given in Eq. (3.13). In order to consider the series of images as a pattern, the measurement operator must be reshaped as follows:

$$S \ : \ \mathcal{O} \times \mathcal{B}_{\theta'} \longrightarrow \mathcal{G}, \quad (o, \cdot) \mapsto \mathcal{S}_g \quad \text{Continuous series of images,}$$

$$S \ : \ \mathcal{O} \times \mathcal{B}' \longrightarrow \mathcal{G}, \quad (o, \cdot) \mapsto \mathcal{S}_g \quad \text{Discrete series of images,}$$

where $\mathcal{G}$ represents a three dimensional space.

The series of images obtained as described in this chapter contain all the necessary information about the surface topography so as to detect and classify defects on it. These series of images constitute the patterns from which the relevant information is collected in form of invariant features. The method to construct such invariants from the given series of images is presented in the following chapter.

5 Extracting invariant features

This chapter presents a method to construct a system of invariant features from a series of images $\mathcal{S}_g$. This system of invariants will then be used to classify $\mathcal{S}_g$ according to the defects it might exhibit. The following issues should be considered in order to design a system of invariant features [Woo96, SM95a]:

- Invariance versus tolerance: Due to the continuous nature of the transformation group $T(\mathcal{T})$ and the discrete nature of computational operations only an approximated invariance can be achieved. Rigorously, a tolerance against the transformations rather than a real invariance can be obtained.

- Discriminability: The system must be able to distinguish between patterns that belong to different equivalence classes.

- Computational complexity: The invariants must be obtained without using an unreasonable amount of time and memory.

- Simplicity and speed of the testing: In practice, the testing process should be as fast and and simple as possible particularly in the case of on-line applications.

- Generalization: The system must be able to generalize to the patterns that are not considered during the training. In particular, the system should be tolerant to noise.

- Locality: Local variations in the patterns must not influence the complete feature space.

Before describing the proposed method to construct invariant features from series of images, the next section presents an introduction to the calculation of invariants by means of integration over the transformation group $T(\mathcal{T})$.

5.1 Invariant features through integration

In order to construct invariants through integration, it is first necessary to describe more precisely the structure of the pattern space $\mathcal{G}$ and that of the transformation group $T(\mathcal{T})$.

Let $\mathcal{G}$ be a topological space. That is, for each pattern $g \in \mathcal{G}$, there exists a family of subsets that are called *neighborhoods* $\mathcal{U}_g \subset \mathcal{G}$ such that [Fle77]:

- Every pattern g has at least one neighborhood $\mathcal{U}_g$,

- $g \in \mathcal{U}_g$ for all $\mathcal{U}_g \subset \mathcal{G}$,

- if $\mathcal{U}_g^1$ and $\mathcal{U}_g^2$ exist for g, then there exists $\mathcal{U}_g^3 \,|\, \mathcal{U}_g^3 \subset (\mathcal{U}_g^1 \cap \mathcal{U}_g^2)$ and

- for $g' \in \mathcal{U}_g$, $\mathcal{U}_{g'}$ can be found such that $\mathcal{U}_{g'} \subset \mathcal{U}_g$.

The family of neighborhoods $\mathcal{U}_g$ is called *topology* of $\mathcal{G}$. Additionally, if every two different patterns g^1 and $g^2 \in \mathcal{G}$ can be separated by disjoint neighborhoods $\mathcal{U}_{g^1} \cap \mathcal{U}_{g^2} = \emptyset$, the topological space $\mathcal{G}$ is separable and receives the name of *Hausdorff space* [Len90]. Furthermore, let $\mathcal{G}$ be locally compact, i.e., each $g \in \mathcal{G}$ has a neighborhood that is closed and bounded. In this case, $\mathcal{G}$ is a *locally compact Hausdorff space*.

In Eq. (3.8), the induced transformation $T(t)$ was defined as a bijective mapping on $\mathcal{G}$. Let now this mapping be more precisely defined to be a continuous, invertible mapping of $\mathcal{G}$ onto itself. This latter definition together with the described structure of $\mathcal{G}$ allows defining the transformation group $T(\mathcal{T})$ as a *topological locally compact group* [Len90].

With this structure of $T(\mathcal{T})$, it is possible to describe each equivalence class in $\mathcal{G}$ as an orbit or trajectory along which all its elements are aligned. For a set of transformation parameters $\mathcal{T} = \{t^e | 1 \le e \le T\}$, the orbit $\mathcal{O}_p$ can be described by the following sequence:

$$\mathcal{O}_p = \langle T(t^e)p, 1 \le e \le T \rangle, \tag{5.1}$$

where again $p \in \mathcal{G}$ is only one pattern selected as prototype. The definitions of orbit $\mathcal{O}_p$ and equivalence class $\mathcal{E}_p$ (Eq. (3.11)) are very similar. However, the concept of orbit implies that $T(\mathcal{T})$ is a topological locally compact group and that a natural and continuous order of the elements can be established [Len90]. Then, an equivalence class $\mathcal{E}_p$ can be described by an orbit $\mathcal{O}_p$ when $T(\mathcal{T})$ is a topological locally compact group. Once the structure of the transformation group has been precisely described, the same must be done with the feature function $f(g)$ (Eq. (3.6)). The support of the function $f(g)$ is defined as the closure of all $g \in \mathcal{G}$, where $f(g) \neq 0$ holds. All continuous, real functions on $\mathcal{G}$ with compact support form a space denoted by $\mathcal{C}_o(\mathcal{G})$. A real function on $\mathcal{C}_o(\mathcal{G})$ is called functional, i.e., a functional $F[f]$ maps a function onto a real number:

$$F : \mathcal{C}_o(\mathcal{G}) \longrightarrow \mathbb{R}. \tag{5.2}$$

Let $T(\mathcal{T}) = \{T(t^e) | 1 \leq e \leq T\}$ be a finite group of transformations and $\langle T(t^e)g | 1 \leq e \leq T \rangle$ be the orbit $\mathcal{O}_g$ that describes the equivalence class $\mathcal{E}_g$. If a second transformation $T(t^k) \in T(\mathcal{T})$, with $k \in [1, T]$, is applied to each element of $\mathcal{O}_g$, this results in a new sequence given by $\langle T(t^k t^e)g | 1 \leq e \leq T \rangle$. However, as $T(\mathcal{T})$ is a group, the correspondence $T(t^e) \longrightarrow T(t^k t^e)$ is a bijection, i.e., $T(t^k t^e)$ is also an element of $T(\mathcal{T})$. Therefore, the second sequence is only a permutation of the first one and both of them describe the same orbit, but beginning from different starting points. If a functional can be found such that $F[f(T(t^e)g)] = F[f(T(t^k t^e)g)]$ for all $k \in [1, T]$ and $g \in \mathcal{G}$, then this functional can be used to construct an invariant feature [Dun73]. In this case, it can be written:

$$\tilde{f}(g) = F[f(g')] \quad \forall\, g' \in \mathcal{O}_g. \tag{5.3}$$

Such a functional is given by the integration over the orbit $\mathcal{O}_g$. Replacing the functional $F[f(g')]$ in Eq. (5.3) by this integration results in:

$$\tilde{f}(g) = \int_{g' \in \mathcal{O}_g} f(g') \, \mathrm{d}\mu(g'). \tag{5.4}$$

As $g' = T(t)g$ for $t \in \mathcal{T}$, the integration can be performed over the transformation group $T(\mathcal{T})$. Now, Eq. (5.4) can be rewritten as:

$$\tilde{f}(g) = \frac{1}{|T(\mathcal{T})|} \int_{T(\mathcal{T})} f(T(t)g) \, \mathrm{d}\mu(t) \quad \forall\, t \in \mathcal{T}, \tag{5.5}$$

where $|T(\mathcal{T})|$ is added to normalize the invariant by the volume of the transformation group:

$$|T(\mathcal{T})| := \int_{T(\mathcal{T})} \mathrm{d}t. \tag{5.6}$$

This approach to obtain invariants by integrating over the transformation group was introduced by Hurwitz [Hur97].

So far, the real world transformation has been induced in the pattern space $\mathcal{G}$ through $\mathrm{T}(\mathcal{T})$. However, computing $\mathrm{T}(t)g$ is generally very demanding, what complicates the calculation of Eq. (5.5). An equivalent and computationally efficient option is to consider $\mathcal{G}$ to remain unchanged while the transformation is induced in the function space $\mathcal{C}_0(\mathcal{G})$. This results in a new transformation $Z(t)$ given by the following mapping:

$$Z : \mathcal{C}_0(\mathcal{G}) \times \mathcal{T} \longrightarrow \mathcal{C}_0(\mathcal{G}) \quad (f(g), t) \mapsto Z(t)\{f(g)\}, \tag{5.7}$$

where $Z(t)\{f(g)\} := f(\mathrm{T}(t^{-1})g)$. In this case, the composition rule defined in Eq. (3.9) can be reshaped to:

$$Z(t^2 t^1)\{f(g)\} = Z(t^2)\{Z(t^1)\{f(g)\}\}. \tag{5.8}$$

Now, Eq. (5.5) can be also written as:

$$\tilde{f}(g) = \frac{1}{|\mathrm{T}(\mathcal{T})|} \int_{\mathrm{T}(\mathcal{T})} Z(t)\{f(g)\} \, d\mu(t) \quad \forall f(g) \in \mathcal{C}_0(\mathcal{G}). \tag{5.9}$$

If the feature $\tilde{f}(g)$ obtained in Eqs. (5.5) and (5.9) is really an invariant, the following equality must hold:

$$Z(t')\{\tilde{f}(g)\} = \tilde{f}(g) \quad \forall \, t' \in \mathcal{T}. \tag{5.10}$$

To prove that the integration leads to such an invariant, $\tilde{f}(g)$ must be replaced by the functional $\mathrm{F}[f(g)]$ and the previous equality (5.10) must still hold:

$$Z(t')\{\mathrm{F}[f(g)]\} = Z(t') \left\{ \frac{1}{|\mathrm{T}(\mathcal{T})|} \int_{\mathrm{T}(\mathcal{T})} Z(t)\{f(g)\} \, d\mu(t) \right\}. \tag{5.11}$$

As $\mathrm{F}[\,\cdot\,]$ and $Z\{\,\cdot\,\}$ are permutable, Eq. (5.11) can be rewritten as follows:

$$Z(t')\{\mathrm{F}[f(g)]\} = \frac{1}{|\mathrm{T}(\mathcal{T})|} \int_{\mathrm{T}(\mathcal{T})} Z(t')\{Z(t)\{f(g)\}\} \, d\mu(t). \tag{5.12}$$

Using the composition rule presented in Eq. (5.8), Eq. (5.12) yields:

$$Z(t')\{\mathrm{F}[f(g)]\} = \frac{1}{|\mathrm{T}(\mathcal{T})|} \int_{\mathrm{T}(\mathcal{T})} Z(t't)\{f(g)\} \, d\mu(t). \tag{5.13}$$

Because $Z(t't)$ is a permutation of $Z(t)$, integrating $Z(t't)\{f(g)\}$ is equal to integrating $Z(t)\{f(g)\}$. Finally, if there exists a measure μ for which $\mu(t't) = \mu(t)$ holds for all $t \in \mathcal{T}$, then Eq. (5.13) results in Eq. (5.10):

$$Z(t')\{F[f(g)]\} = \frac{1}{|T(\mathcal{T})|} \int_{T(\mathcal{T})} Z(t)f(g)\,\mathrm{d}\mu(t) = \tilde{f}(g). \qquad (5.14)$$

The existence of such a measure μ was rigorously proved by Haar in [Haa33]. Haar demonstrated that there exists a non-zero regular Borel measure μ which is left/right invariant, if $T(\mathcal{T})$ is a topological locally compact group [Haa33]. As a consequence, Eq. (5.5) is known as Haar integral, while μ is called Haar measure. The resulting invariant $\tilde{f}(g)$ is called integral invariant.

5.1.1 System of invariants

In the calculation of invariants by integration, the real function $f(\,\cdot\,)$ is called *kernel function*. In general, this function depends not only on the pattern g, but also on a parameter vector $\mathbf{w}^l$ with $l \in \{1,\dots,L\}$. Thus, the kernel function can be defined as follows:

$$f : \mathcal{G} \times \mathcal{W} \longrightarrow \mathcal{F}; \quad (g, \mathbf{w}^l) \mapsto f^l(g) := f(g, \mathbf{w}^l), \qquad (5.15)$$

where $\mathcal{W} = \{\mathbf{w}^l \,|\, l \in \mathbb{N}\}$ is the set of all possible function parameter vectors $\mathbf{w}^l$ and $\mathcal{F} = \mathbb{R}$. In this case, the set $\mathcal{W}$ can also be included in the definition of functional Eq. (5.2):

$$\mathrm{F} : \mathcal{C}_o(\mathcal{G} \times \mathcal{W}) \longrightarrow \mathbb{R}.$$

Now, the integral invariant $\tilde{f}^l(g) := \tilde{f}(g, \mathbf{w}^l)$ depends also on $\mathbf{w}^l$ and the Haar integral from Eq. (5.5) can be rewritten as follows:

$$\tilde{f}^l(g) = \frac{1}{|T(\mathcal{T})|} \int_{T(\mathcal{T})} f^l(T(t)g)\,\mathrm{d}t, \qquad (5.16)$$

where the special case $\mu(t) := t$ was considered. By selecting a large enough number L of function parameter vectors $\mathbf{w}^l$, a complete system of invariants $\tilde{\mathcal{F}}(g) = \left\langle \tilde{f}^l(g), l \in \{1,\dots,L\} \right\rangle$ can be reached as defined in Eq. (3.12) [Dun73]. The construction of the complete system of invariants $\tilde{\mathcal{F}}(g)$ from the individual invariants $\tilde{f}(g)$ can be described by the following mapping:

$$\tilde{\mathcal{F}}(g) : \mathcal{F} \longrightarrow \mathcal{F}^L.$$

Figure 5.1 illustrates the construction of a complete system of invariants for three non-equivalent patterns g^1, g^2 and g^3 using two different kernel function parameter vectors $\mathbf{w}^1$ and $\mathbf{w}^2$.

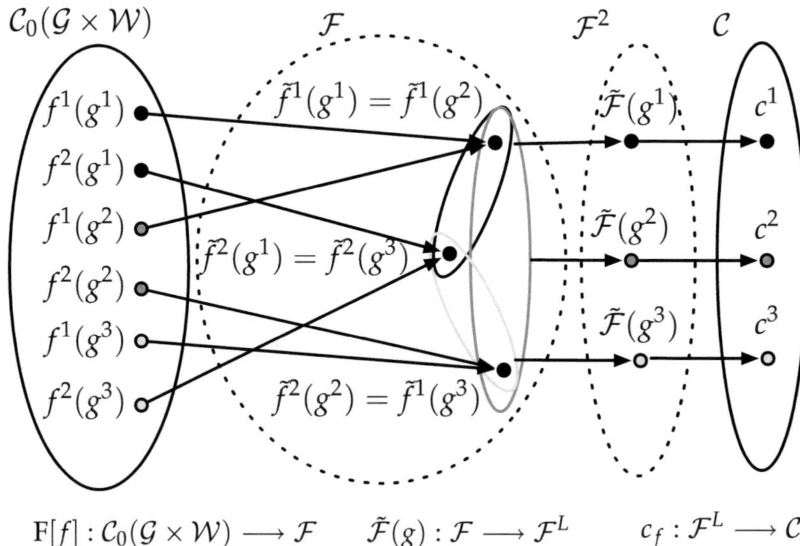

Figure 5.1: Construction of complete invariants for $L = 2$. In the example, $\tilde{f}^1(g^1) = \tilde{f}^1(g^2)$ and $\tilde{f}^2(g^1) = \tilde{f}^2(g^3)$, i.e., the invariant $\tilde{f}^1(\cdot)$ alone cannot distinguish between the classes c_1 and c_2, while the invariant $\tilde{f}^2(\cdot)$ alone cannot differentiate between the classes c_1 and c_3. Only the complete system $\tilde{\mathcal{F}}(g) = \langle \tilde{f}^1(g), \tilde{f}^2(g) \rangle$ can recognize the three classes.

5.2 Extending the Haar integral for series of images

The Haar integral presented in Eq. (5.16) allows obtaining a feature $\tilde{f}^l(g)$ from a pattern g, which is invariant under the transformation group $T(\mathcal{T})$. On the other hand, in Chapter 4, patterns representing varnished surfaces were defined as series of images $\mathcal{S}_g$. Hence, to obtain invariants that characterize this kind of surfaces, g must be replaced by $\mathcal{S}_g$ in Eq. (5.16). Further, as the integral requires continuous variables, the definition of continuous series of images given in Eq. (4.1) must be used:

$$\tilde{f}^l(\mathcal{S}_g) = \frac{1}{|T(\mathcal{T})|} \int_{T(\mathcal{T})} f^l(T(t)\mathcal{S}_g)\, dt, \tag{5.17}$$

where $f^l(\mathcal{S}_g)$ can be defined as follows for a three dimensional space $\mathcal{G}$:

$$f : \mathcal{G} \times \mathcal{W} \longrightarrow \mathcal{F} = \mathbb{R}; \quad (\mathcal{S}_g, \mathbf{w}^l) \mapsto f^l(\mathcal{S}_g). \tag{5.18}$$

The cyclical 2D Euclidean motion was introduced as the transformation group of interest (Eq. (3.17)). Introducing this group into Eq. (5.17) results in:

$$\tilde{f}^l(\mathcal{S}_g) = \frac{1}{|M(\mathcal{M})|} \int_{M(\mathcal{M})} f^l(M(\tau_x, \tau_y, \varphi)\mathcal{S}_g)\, d(\tau_x, \tau_y, \varphi),$$

or:

$$\tilde{f}^l(\mathcal{S}_g) = \frac{1}{|M(\mathcal{M})|} \int_{\tau_x} \int_{\tau_y} \int_{\varphi} f^l(M(\tau_x, \tau_y, \varphi)\mathcal{S}_g)\, d\tau_x\, d\tau_y\, d\varphi. \tag{5.19}$$

However, in practice, the images are digital and the acquisition of the series of images is not continuous. As a consequence, $\mathcal{S}_g$ is a discrete series of images and must be defined as in Eq. (4.3). Additionally, the transformation group $M(\mathcal{M})$ must be discretized as described in Eq. (3.18). In order to consider the discrete nature of pattern and transformation group, the Haar integral is replaced by summations as follows:

$$\tilde{f}^l(\mathcal{S}_g) = \frac{1}{MNK} \sum_{i=0}^{M-1} \sum_{j=0}^{N-1} \sum_{k=0}^{K-1} f^l(M_{ijk}\mathcal{S}_g). \tag{5.20}$$

If the transformation is induced in the function space $\mathcal{C}_0(\mathcal{G})$ as proposed in Eq. (5.9), the previous summation can also be written as:

$$\tilde{f}^l(\mathcal{S}_g) = \frac{1}{MNK} \sum_{i=0}^{M-1} \sum_{j=0}^{N-1} \sum_{k=0}^{K-1} Z_{ijk}\{f^l(\mathcal{S}_g)\}. \tag{5.21}$$

5.2.1 2D Euclidean motion for series of images

As discussed, Eq. (3.18) describes the cyclical 2D Euclidean motion transformation on a discrete image g_{mn}. However, this equation does not consider that the surface has a relief and that it is imaged under directional light. The relative position between the surface relief and the light source changes as the surface rotates, which causes a new arrangement of light and shadow on the surface. This results in local intensity changes that are captured in the images [CW00, WC03] and are not considered in Eq. (3.18). This section is concerned with the combination of directional illumination with 2D Euclidean motion in oder to determine how it affects the series of images $\mathcal{S}_g$.

The transformed discrete series of images $\mathcal{S}'_g := M_{ijk}\,\mathcal{S}_g$ is denoted by:

$$\mathcal{S}'_g = \langle g'_{mnb}, b \in \{0, 1, \dots, B-1\}\rangle.$$

The transformation of the series of images according to the 2D Euclidean motion can be performed in two steps. First, each image of the series is individually transformed according to Eq. (3.18). Thus, the rotation and translation of each image of the series g_{mnb} results in a transformed image denoted by $g_{m'n'b} := M_{ijk}\,g_{mnb}$, where:

$$\begin{pmatrix} m' \\ n' \end{pmatrix} = \begin{pmatrix} (m\cos k_\varphi - n\sin k_\varphi + i)\bmod M \\ (m\sin k_\varphi + n\cos k_\varphi + j)\bmod N \end{pmatrix}. \tag{5.22}$$

The second step of the transformation considers the intensity changes caused by the interaction of the surface relief and the directional illumination. Let this effect be illustrated through an example. Figure 5.2 (a) shows a top view of a defect under diffuse illumination. (The shape of this defect is fictitious and has been selected with the only purpose of illustrating the problem.) Additionally, four different positions are indicated around the defect using different symbols: a circle, a rhombus, a triangle, and a pentagon. Figure 5.2 (b) shows the resulting image after an anti-clockwise rotation of the surface in $270°$. In this transformed image, the coordinates of these symbols must also be rotated in $270°$ so as to relocate them in the same positions with respect to the defect. The symbols in the transformed image have not only the same positions with respect to the defect, but, as diffuse illumination has been considered, the intensity values at these positions are also the same as in the original image. Thus, in this case, Eq. (3.18) gives a full description of the transformation.

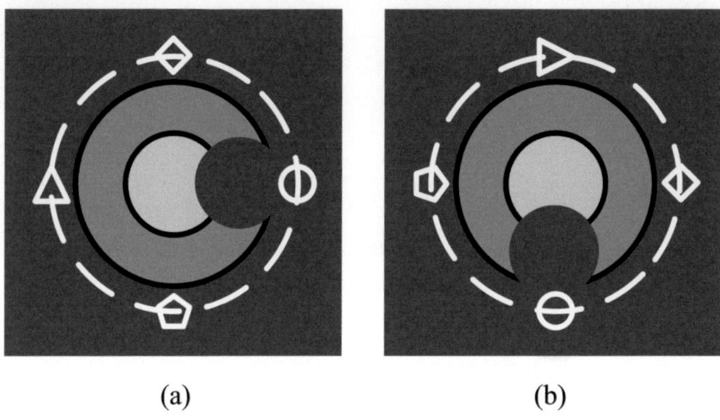

(a)   (b)

Figure 5.2: Top view of a simulated defect.

Now, Figure 5.3 (left) illustrates the same previous defect, but this time under directional illumination. In addition, a complete series of images $\mathcal{S}_g$ with $B = 4$ ($\Delta\varphi = 90°$) is taken into account. The same positions indicated in Fig. 5.2 (a) are located in the series of images. However, in this case, the positions of symbols must be considered not only with respect to the defect, but also with respect to the illumination source. In order to keep the same relative position between each symbol and the illuminant, the symbols must be located in different images of the series. In this example, they have always been placed in front of the light source. Figure 5.3 (right) shows the series of images that results from rotating the surface by $270°$. As previously, to relocate the given symbols in the same positions with respect to the defect, their original coordinates must also be rotated by $270°$. However, the intensity values at these positions do not coincide with the ones in the original series of images. In order to obtain the original intensity values, the symbols must also be relocated with respect to the illuminant. This can be achieved by cyclically shifting symbols by $270°$ ($b\Delta\omega$) along the series of images. In conclusion, the first step (rotation) relocates each symbol in the same position with respect to the defect, whereas the second step (cyclic shift along the series of images) repositions each symbol with respect to the illuminant. Then, considering Eq. (5.22) and the cyclic shift along b, the complete transformation of $\mathcal{S}_g$ according to $M(\mathcal{M})$ is given by:

$$\mathcal{S}'_g = \left\langle g'_{mnb} := g_{m'n'b'},\ b' \in \{0, 1, \ldots, B - 1\}\right\rangle, \tag{5.23}$$

where

$$b' = (b + k) \bmod B. \tag{5.24}$$

From Eq. (5.24), it can be concluded that for discrete series of images the resolution $1/\Delta\omega$ used during the image acquisition (Eq. (4.2)) limits the resolution of the rotation transformation $\Delta\varphi$ to:

$$\Delta\varphi = \nu\,\Delta\omega \quad \nu \in \mathbb{N}. \tag{5.25}$$

5.2.2 Kernel function

The function $f(g)$ should extract the relevant information from the patterns (Eq. (3.6)). In Section 5.1, this function was defined as a real one with compact support. Moreover, Eq. (5.15) and Eq. (5.18) extend the definition of $f(g)$ to incorporate a parameter vector $\mathbf{w}^l$ and to consider series of images respectively. Additionally, by introducing this function into the Haar integral, it acquired the name of kernel function.

In this section, a concrete kernel function for the inspection of varnished surfaces is proposed. To design an appropriate kernel function, two aspects related to the surface characteristics must be contemplated:

- Some information about the existence and type of topographic defects is contained in intensity changes within 2D neighborhoods in each image of the series. As a consequence, the kernel function should be able to take this into account.

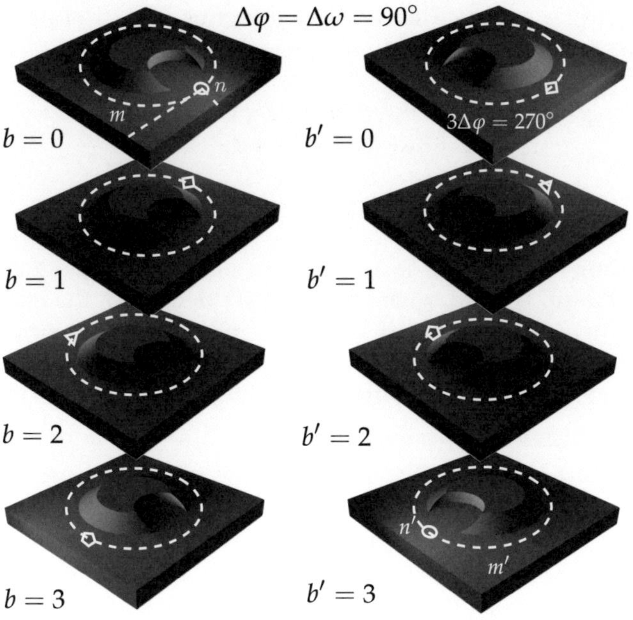

Figure 5.3: Rotation transformation for a series of images with $B = 4$ ($\Delta\omega = \Delta\varphi = 90°$), and $k = 3$ ($k\Delta\varphi = 270°$).

- Intensity variations along the series of images (third dimension given by b) enclose additional information about the existence and type of defects. The kernel function should also collect this information.

The proposed kernel function is a vectorial function based on the LBP operator (Section 2.1.1) and is denoted by:

$$\mathbf{f}^l(\mathcal{S}_g) = (f_l^1(\mathcal{S}_g),\ldots,f_l^{Q_l}(\mathcal{S}_g)), \tag{5.26}$$

with $Q_l \in \mathbb{N}$ and $\mathbf{w}^l = (r_{1,l}, r_{2,l}, \alpha_l, \beta_l, a_l, \Delta\rho_l)$. The q-th element of $\mathbf{f}^l(\mathcal{S}_g)$ is defined as follows:

$$f_l^q(\mathcal{S}_g) = \left| g_{\mathbf{u}_l^q} - g_{\mathbf{v}_l^q} \right|, \tag{5.27}$$

where $q \in \{1, \ldots, Q_l\}$. The coordinate vectors $((m,n)^{\mathrm{T}}; b)$ denoted by $\mathbf{u}_l^q$ and $\mathbf{u}_l^q$ are given by the following expression:

$$
\begin{aligned}
\mathbf{u}_l^q &= \left(\begin{bmatrix} r_{1,l} \cos(\alpha_l + q\,\Delta\rho_l) \\ -r_{1,l} \sin(\alpha_l + q\,\Delta\rho_l) \end{bmatrix}; 0 \right), \\
\mathbf{v}_l^q &= \left(\begin{bmatrix} r_{2,l} \cos(\beta_l + q\,\Delta\rho_l) \\ -r_{2,l} \sin(\beta_l + q\,\Delta\rho_l) \end{bmatrix}; a_l \right).
\end{aligned}
\tag{5.28}
$$

According to Eq. (5.28), two circular neighborhoods centered on $m = n = 0$ with radii $r_{1,l}$ and $r_{2,l}$ are respectively defined in the images g_{mn0} and g_{mna_l} of $\mathcal{S}_g$, where $a_l \in \{0, \ldots, B-1\}$ indicates one particular image of the series. Both circumferences are sampled with a frequency given by the angle $\Delta\rho_l$. This sampling results in $Q_l = \frac{2\pi}{\Delta\rho_l}$ points per neighborhood, which are respectively addressed through the vectors $\mathbf{u}_l^q$ and $\mathbf{v}_l^q$. For a given value of q, the corresponding function element $f_l^q(\mathcal{S}_g)$ is obtained taking the absolute value of the difference between intensities addressed by the vectors $\mathbf{u}_l^q$ and $\mathbf{v}_l^q$ (Eq. (5.27)). Figure 5.4 illustrates the kernel function for $\mathbf{w}^l = (1, 2, 45°, 90°, 2, 180°)$. In this figure, the intensity values given by $\mathbf{u}_l^q$ and $\mathbf{v}_l^q$, involved in the calculation of each element $f_l^q(\mathcal{S}_g) \in \mathbf{f}^l(\mathcal{S}_g)$, are linked through blue segments.

As mentioned before, the Haar integral can be calculated more efficiently if the transformation is induced in the function space and not in the pattern space (Eq. (5.7)). Hence, the cyclical 2D Euclidean motion transformation should be applied to the defined kernel function and not to the series of images as in Eq. (5.23). Let $\mathcal{C}_0'(\mathcal{G}) \subset \mathcal{C}_0(\mathcal{G})$ be the subset of the real vectorial functions defined in Eq. (5.26). The transformation of the kernel function, according to $\mathrm{M}(\mathcal{M})$, can be described by the following mapping:

$$
\mathrm{Z} : \mathcal{C}_0'(\mathcal{G}) \times \mathcal{M} \longrightarrow \mathcal{C}_0'(\mathcal{G}); \quad (\mathbf{f}^l(\mathcal{S}_g), i, j, k) \mapsto \mathrm{Z}_{ijk}\{\mathbf{f}^l(\mathcal{S}_g)\}.
$$

The transformed kernel function results in:

$$
\mathbf{f}_{ijk}^l(\mathcal{S}_g) := \mathrm{Z}_{ijk}\{\mathbf{f}^l(\mathcal{S}_g)\} = (f_{lijk}^1(\mathcal{S}_g), \ldots, f_{lijk}^{Q_l}(\mathcal{S}_g)),
$$

with its q-th element defined as:

$$
f_{lijk}^q(\mathcal{S}_g) = \left| g_{\mathbf{u}_{lijk}^q} - g_{\mathbf{v}_{lijk}^q} \right|.
\tag{5.29}
$$

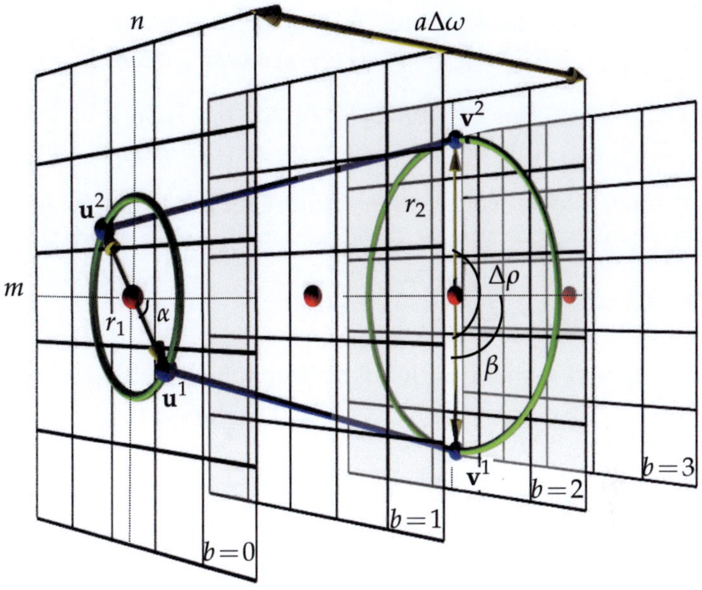

Figure 5.4: Kernel function $\mathbf{f}^l(\mathcal{S}_g)$ for a series of images with $B = 4\,(\Delta\omega = 90°)$. Parameters: $r_{1,l} = 1$, $r_{2,l} = 2$, $\alpha_l = 45°$, $\beta_l = 90°$, $a_l = 2$ and $\Delta\rho_l = 180°\,(Q_l = 2)$. The local center of coordinates is indicated by m and n. The blue segments between $\mathbf{u}$ and $\mathbf{v}$ represent the absolute value of the difference between the addressed intensities $\left| g_{\mathbf{u}_l^q} - g_{\mathbf{v}_l^q} \right|$. For ease of exposition the index l has been suppressed from this figure.

From Eqs. (5.28), (5.22) and (5.23), the following expressions for the vectors $\mathbf{u}_{lijk}^q$ and $\mathbf{v}_{lijk}^q$ can be obtained:

$$\mathbf{u}_{lijk}^q = \left(\begin{bmatrix} u_{lik}^q \bmod M \\ u_{ljk}^q \bmod N \end{bmatrix} ; (0+k) \bmod B \right) ,$$

$$\mathbf{v}_{lijk}^q = \left(\begin{bmatrix} v_{lik}^q \bmod M \\ v_{ljk}^q \bmod N \end{bmatrix} ; (a_l+k) \bmod B \right) , \qquad (5.30)$$

$$\begin{bmatrix} u^q_{lik} \\ u^q_{ljk} \end{bmatrix} = \begin{bmatrix} \cos k_\varphi & \sin k_\varphi \\ \sin k_\varphi & \cos k_\varphi \end{bmatrix} \begin{bmatrix} r_{1,l}\cos(\alpha_l + q\,\Delta\rho_l) \\ -r_{1,l}\sin(\alpha_l + q\,\Delta\rho_l) \end{bmatrix} + \begin{bmatrix} i \\ j \end{bmatrix},$$

$$\begin{bmatrix} v^q_{lik} \\ v^q_{ljk} \end{bmatrix} = \begin{bmatrix} \cos k_\varphi & \sin k_\varphi \\ \sin k_\varphi & \cos k_\varphi \end{bmatrix} \begin{bmatrix} r_{2,l}\cos(\beta_l + q\,\Delta\rho_l) \\ -r_{2,l}\sin(\beta_l + q\,\Delta\rho_l) \end{bmatrix} + \begin{bmatrix} i \\ j \end{bmatrix}.$$

Now, in order to obtain an invariant feature for this kernel function, the transformed function must be introduced in Eq. (5.21), which results in:

$$\tilde{\mathbf{f}}^l(\mathcal{S}_g) = \frac{1}{MNK} \sum_{i=0}^{M-1} \sum_{j=0}^{N-1} \sum_{k=0}^{K-1} \mathbf{f}^l_{ijk}(\mathcal{S}_g). \tag{5.31}$$

As the proposed kernel function is vectorial, the resulting invariant is also a vector $\tilde{\mathbf{f}}^l(\mathcal{S}_g) = (\tilde{f}^1_l(\mathcal{S}_g), \ldots, \tilde{f}^{Q_l}_l(\mathcal{S}_g))$, whose q-th element is:

$$\tilde{f}^q_l(\mathcal{S}_g) = \frac{1}{MNK} \sum_{i=0}^{M-1} \sum_{j=0}^{N-1} \sum_{k=0}^{K-1} f^q_{lijk}(\mathcal{S}_g). \tag{5.32}$$

5.3 Construction of invariant features

This section describes how to design the invariant vector $\tilde{\mathbf{f}}^l(\mathcal{S}_g)$ based on Eq. (5.31). In order to improve the discriminability properties of $\tilde{\mathbf{f}}^l(\mathcal{S}_g)$, two different methods are applied to obtain invariance against rotation and translation. The following sections present a description of both proposed methods.

5.3.1 Invariance against rotation

Considering in Eq. (5.31) only the summation over k results in:

$$\tilde{\mathbf{f}}^l_{ij}(\mathcal{S}_g) = \frac{1}{K} \sum_{k=0}^{K-1} \mathbf{f}^l_{ijk}(\mathcal{S}_g), \tag{5.33}$$

where $\tilde{\mathbf{f}}^l_{ij}(\mathcal{S}_g) = (\tilde{f}^1_{lij}(\mathcal{S}_g), \ldots, \tilde{f}^{Q_l}_{lij}(\mathcal{S}_g))$ is a Q_l-dimensional vector of invariants against rotation, whose q-th element can be defined from Eq. (5.32) as follows:

$$\tilde{f}^q_{lij}(\mathcal{S}_g) = \frac{1}{K} \sum_{k=0}^{K-1} f^q_{lijk}(\mathcal{S}_g). \tag{5.34}$$

Each element $f_{lijk}^{q}(\mathcal{S}_g)$ is calculated according to Eq. (5.29). Then, Eq. (5.34) can also be written as:

$$\tilde{f}_{lij}^{q}(\mathcal{S}_g) = \frac{1}{K} \sum_{k=0}^{K-1} \left| g_{\mathbf{u}_{lijk}^{q}} - g_{\mathbf{v}_{lijk}^{q}} \right|.$$ (5.35)

For each value of k in Eq. (5.35), the coordinates $\mathbf{u}_{lijk}^{q}$ and $\mathbf{v}_{lijk}^{q}$ are rotated in $k_{\varphi} := k\Delta\varphi$ and shifted along the series of images in $k \bmod B$ (Eq. (5.30)). Figure 5.5 illustrates the calculation of Eq. (5.33) for each value of k with an example. A series of images with $B = 4$ and the kernel function $\mathbf{f}^{l}(\mathcal{S}_g)$ illustrated in Fig. 5.4 are considered. The resolution of the rotation transformation is fixed to $\Delta\varphi = 90°$, which results in $k \in \{0, 1, 2, 3\}$. The selected kernel function has $Q_l = 2$; then, its transformed version can be written as $\mathbf{f}_{ijk}^{l}(\mathcal{S}_g) = (f_{lijk}^{1}(\mathcal{S}_g), f_{lijk}^{2}(\mathcal{S}_g))$. Figure 5.5(a) and Fig. 5.5(b) illustrate the variations of $f_{lijk}^{1}(\mathcal{S}_g)$ and $f_{lijk}^{2}(\mathcal{S}_g)$ respectively for the different values of k.

5.3.2 Invariance against translation: fuzzy histograms

For a complete invariance against 2D Euclidean motion, it is necessary also to consider the translation transformation. For this purpose, it is clearly possible to perform the summation over i and j in Eq. (5.31). In this case, Eq. (5.33) should be introduced into Eq. (5.31), which results in:

$$\tilde{\mathbf{f}}^{l}(\mathcal{S}_g) = \frac{1}{MN} \sum_{i=0}^{M-1} \sum_{j=0}^{N-1} \tilde{\mathbf{f}}_{ij}^{l}(\mathcal{S}_g),$$

where each element of $\tilde{\mathbf{f}}^{l}(\mathcal{S}_g)$ is defined as:

$$\tilde{f}_{l}^{q}(\mathcal{S}_g) = \frac{1}{MN} \sum_{i=0}^{M-1} \sum_{j=0}^{N-1} \tilde{f}_{lij}^{q}(\mathcal{S}_g).$$

Alternatively, these summations over i and j can be replaced by histograms, which are intrinsically invariant against translation. Unlike summing over i and j, histograms have the advantage of reducing loss of information. Thus, features generated based on histograms can better

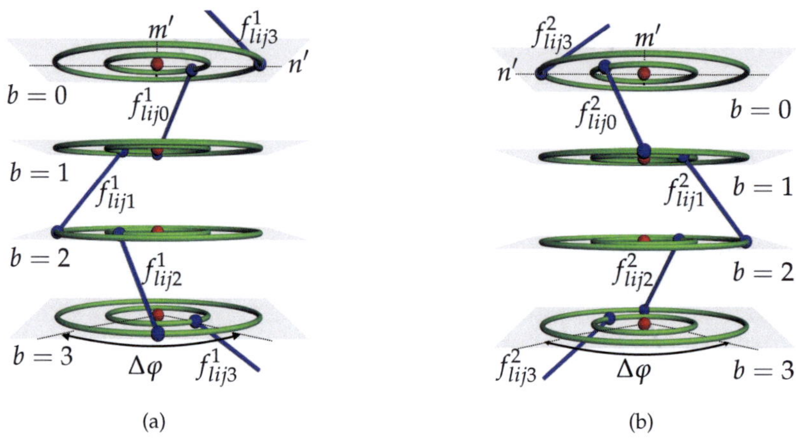

Figure 5.5: Extraction of rotation invariant features $\tilde{f}^q_{lij}$ for the kernel function elements (a) $f^1_{lijk}(\mathcal{S}_g)$ and (b) $f^2_{lijk}(\mathcal{S}_g)$ illustrated in Fig. 5.4. The blue segments link the points $\mathbf{u}^q_{lijk}$ and $\mathbf{v}^q_{lijk}$ used to calculate $f^q_{lijk}(\mathcal{S}_g)$ for each value of k in the summation of Eq. (5.34). The values of i and j are fixed and the resulting coordinates are denoted by m' and n'. For ease of exposition the argument $(\mathcal{S}_g)$ has been suppressed from this figure.

characterize the different classes. The combination of integral invariants and feature histograms is extensively used in pattern recognition [Sig02, HB04, SIB05, ZKL08] (see Section 2.3).

In this work, histograms are built from the invariants against rotation $\tilde{f}^q_{lij}(\mathcal{S}_g) \in \tilde{\mathbf{f}}^l_{ij}(\mathcal{S}_g)$ (Eq. (5.34)). However, traditional histograms present the disadvantage of being discontinuous at bin boundaries. Therefore, small variations in $\tilde{f}^q_{lij}(\mathcal{S}_g)$ can cause big changes in the histograms, which results in features that are not robust against noise or local variations. This problem can be solved by constructing fuzzy histograms, which avoid this discontinuous assignment of bins. A complete analysis of fuzzy histograms can be found in [Sig02].

The construction of fuzzy histograms begins with the generation of matrices of rotation invariants. For a given value of q, an $M \times N$ matrix $\mathbf{H}^q_l(\mathcal{S}_g)$, whose elements are denoted by (i, j), can be constructed with the invariants $\tilde{f}^q_{lij}(\mathcal{S}_g)$ (Eq. (5.34)). Now, a sequence $\mathcal{H}^l(\mathcal{S}_g)$ of matrices $\mathbf{H}^q_l(\mathcal{S}_g)$ can be obtained changing the value of q from 1 to Q_l:

$$\mathcal{H}^l(\mathcal{S}_g) = \left\langle \mathbf{H}^q_l(\mathcal{S}_g); q \in \{1, \ldots, Q_l\} \right\rangle. \tag{5.36}$$

Finally, from each matrix $\mathbf{H}^q_l(\mathcal{S}_g)$, a fuzzy histogram $\mathcal{H}^q_{ld}(\mathcal{S}_g)$ is built, where $d \in \{1, \ldots, D\}$ represents the histogram bins. By selecting a maximal number of histogram bins D, their range can be calculated as $\Delta = \frac{\hat{f}}{D}$, where $\hat{f}$ is the maximal possible value of $\tilde{f}^q_{lij}(\mathcal{S}_g)$. From Eq. (5.29) and Eq. (5.34), it can be observed that $\hat{f}$ is given by the maximal gray level of the used image dynamic range. Denoting $\Delta_d := d \cdot \Delta$, the value of the fuzzy histogram bin d' is calculated as follows:

$$\mathcal{H}^q_{ld'}(\mathcal{S}_g) = \begin{cases} \displaystyle\sum_{\Delta_0 \leq \tilde{f} < \Delta_1} 1 - \left(\frac{\tilde{f} - \Delta_0}{\Delta}\right) & d' = 0, \\[2em] \displaystyle\sum_{\Delta_{d'-1} \leq \tilde{f} \leq \Delta_{d'}} \frac{\tilde{f} - \Delta_{d'-1}}{\Delta} + \sum_{\Delta_{d'} \leq \tilde{f} \leq \Delta_{d'+1}} 1 - \left(\frac{\tilde{f} - \Delta_{d'}}{\Delta}\right) & d' \in (0,D), \\[2em] \displaystyle\sum_{\Delta_{D-1} \leq \tilde{f} < \Delta_D} \frac{\tilde{f} - \Delta_{D-1}}{\Delta} & d' = D, \end{cases}$$

where $\tilde{f}^q_{lij}(\mathcal{S}_g)$ was written as $\tilde{f}$ to simplify the notation.

By calculating $\mathcal{H}^q_{ld}(\mathcal{S}_g)$ for each $\mathbf{H}^q_l(\mathcal{S}_g) \in \mathcal{H}^l(\mathcal{S}_g)$, a sequence of histograms is obtained, which together represent the proposed invariant feature:

$$\tilde{\mathbf{f}}^l(\mathcal{S}_g) = \left\langle \mathcal{H}^q_{ld}(\mathcal{S}_g); q \in \{1, \ldots, Q_l\} \right\rangle. \tag{5.37}$$

In conclusion, invariance against rotation has been achieved through the summation over the transformation group, while invariance against translation has been obtained by constructing fuzzy histograms. This way, $\tilde{\mathbf{f}}^l(\mathcal{S}_g)$ contains both gray-level and structural information about $\mathcal{S}_g$.

In order to achieve a complete system of invariant features, a subset $\mathcal{W}'$ of parameter vectors $\mathbf{w}^l$ must be selected ($\mathcal{W}' \subset \mathcal{W}$) for a big enough $|\mathcal{W}'| = L$. The complete system of invariants is denoted by:

$$\tilde{\mathcal{F}}(\mathcal{S}_g) = \left\langle \tilde{\mathbf{f}}^1(\mathcal{S}_g), \ldots, \tilde{\mathbf{f}}^L(\mathcal{S}_g) \right\rangle. \tag{5.38}$$

5.3.3 Improving class separability

When detecting and classifying varnish defects, some changes can be incorporated to the construction of invariants in order to improve the separability between classes. First, the real dynamic range of the kernel function elements $f_l^q(\mathcal{S}_g)$ should be increased.

According to Eq. (5.27), the dynamic range of the images $g_{mnb} \in \mathcal{S}_g$ determines the dynamic range of $f_l^q(\mathcal{S}_g)$. For example, if gray levels of the series of images are coded with eight bits, then their dynamic range is given by $2^8 = 256$ (from 0 to 255). As the elements of the kernel function $f_l^q(\mathcal{S}_g)$ are obtained from the absolute difference between gray levels of $\mathcal{S}_g$, their values can also vary between 0 and 255. The same is valid for the transformed kernel function elements, i.e., for $f_{lijk}^q(\mathcal{S}_g)$. However, in practice, gray levels present small variations inside series of images of varnished wood surfaces, even if there exist defects. This means that the dynamic range is actually smaller than in theory. Further, this reduced dynamic range is preserved during the calculation of the rotation invariants $\tilde{f}_{lij}^q(\mathcal{S}_g)$ (Eq. (5.35)). These rotation invariants $\tilde{f}_{lij}^q(\mathcal{S}_g)$ are then used to construct the histograms $\mathcal{H}_{ld}^q(\mathcal{S}_g)$, whose sequence generates $\tilde{\mathbf{f}}^l(\mathcal{S}_g)$ (Eq. (5.37)). Thus, if the number of bins D of $\mathcal{H}_{ld}^q(\mathcal{S}_g)$ is not chosen to be big enough, there could be an irreversible loss of relevant information. This latter can result in features that cannot distinguish between classes. On the other hand, a big D can produce histograms with many empty bins, which will also negatively affect the classification.

Figure 5.6 exemplifies this problem by plotting the values $\tilde{f}_{lij}^q(\mathcal{S}_g)$ obtained from two series of images. The presented series of images with $B = 4$ belong to different classes: $\mathcal{S}_{g^1}$ shows a surface without any defect and $\mathcal{S}_{g^2}$ shows a surface with a fissure. The used kernel function parameters are given by $\mathbf{w}^l = (0, 3, 0°, 0°, 1, 90°)$ (i.e., $Q_l = 4$). From the resulting values $\tilde{f}_{lij}^q(\mathcal{S}_g)$, histograms $\mathcal{H}_{ld}^q(\mathcal{S}_g)$ are then calculated. In this case, if D is set to be 5, the fuzzy histograms $\mathcal{H}_{ld}^q(\mathcal{S}_{g^1})$ and $\mathcal{H}_{ld}^q(\mathcal{S}_{g^2})$ are almost identical (traditional histograms are identical for $D \leq 5$). In conclusion, the resulting features $\tilde{\mathbf{f}}^l(\mathcal{S}_{g^1})$ and $\tilde{\mathbf{f}}^l(\mathcal{S}_{g^2})$ are not able to reliably distinguish between these two different classes (no defect and fissure).

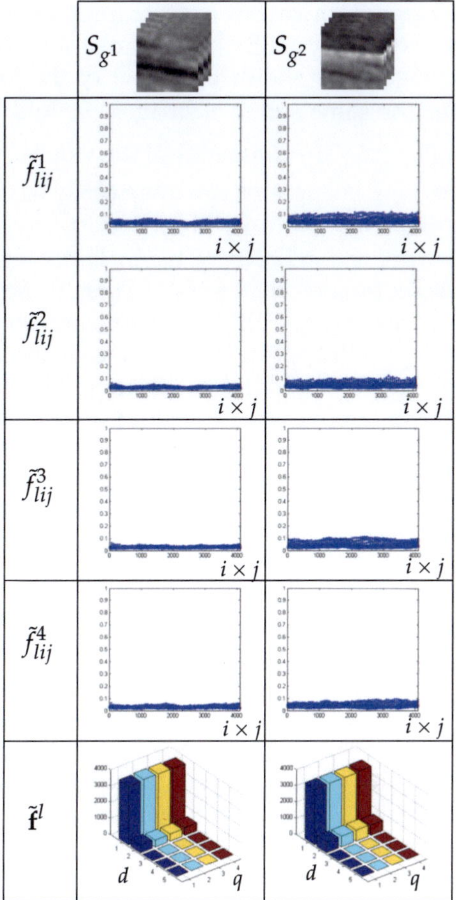

Figure 5.6: $\tilde{f}^{q}_{lij}(\mathcal{S}_g)$ obtained by $\mathbf{f}^l(\mathcal{S}_g)$ with $\mathbf{w}^l = (0,3,0°,0°,1,90°)$ (Eq. (5.27)) for $\mathcal{S}_{g^1}$ (intact surface) and $\mathcal{S}_{g^2}$ (fissure), and the resulting features $\tilde{\mathbf{f}}^l(\mathcal{S}_g)$.

A modified kernel function $f^l(\mathcal{S}_g)$ is proposed to solve this problem, whose q-th element $f^q_l(\mathcal{S}_g)$ is defined as follows:

$$f^q_l(\mathcal{S}_g) = \log_2 \left| g_{\mathbf{u}^q_l} - g_{\mathbf{v}^q_l} + 1 \right| . \tag{5.39}$$

The use of the logarithm causes a decompression of $f_l^q(\mathcal{S}_g)$ in the range of interest and consequently a wider distribution of values into histogram bins. Figure 5.7 shows the effect of the logarithm on the values $\tilde{f}_{lij}^q(\mathcal{S}_g)$ for the same series of images $\mathcal{S}_{g^1}$ and $\mathcal{S}_{g^2}$ and for the same $\mathbf{w}^l$ used in Fig. 5.6. The dynamics of the values $\tilde{f}_{lij}^q(\mathcal{S}_g)$ is clearly larger and the resulting histograms can more easily be differentiated for $D = 5$. As a consequence, features $\tilde{\mathbf{f}}^l(\mathcal{S}_{g^1})$ and $\tilde{\mathbf{f}}^l(\mathcal{S}_{g^2})$ obtained using the kernel function defined in Eq. (5.39) have a better separability for the two presented classes (no defect and fissure) than those obtained by the kernel function defined in Eq. (5.27). In the remainder of this thesis, the kernel function $\mathbf{f}^l(\mathcal{S}_g)$ will apply a logarithm as defined in Eq. (5.39).

In addition, the small size of defects with respect to the surface makes the classification even more difficult. From the used illumination configuration and surface topography (Chapter 4), it can be concluded that high values of $\tilde{f}_{lij}^q(\mathcal{S}_g)$ are most probably generated by defects, whereas low values are typically due to planar regions of the surface. High values of $\tilde{f}_{lij}^q(\mathcal{S}_g)$ increment the last bins of $\mathcal{H}_{ld}^q(\mathcal{S}_g)$, while low values are concentrated in its first bins. The use of a logarithm in the kernel function improves the distance between values $\tilde{f}_{lij}^q(\mathcal{S}_g)$ in the range of interest. However, although the inspected surface shows a defect, the number of $\tilde{f}_{lij}^q(\mathcal{S}_g)$ with high values is still significantly smaller than the number of $\tilde{f}_{lij}^q(\mathcal{S}_g)$ with low values. This is because there is a big difference between the size of defects and the inspected surface. Thus, the last bins of $\mathcal{H}_{ld}^q(\mathcal{S}_g)$ have almost no weight in relation to its first bins. As a consequence, the last bins of $\mathcal{H}_{ld}^q(\mathcal{S}_g)$ have almost no influence during the classification process. In order to improve this situation, an exponential weight function z^d is used to modify histograms $\mathcal{H}_{ld}^q(\mathcal{S}_g)$, where $z \in \mathbb{N}$ and $d \in \{1,\ldots,D\}$ represents the histogram bins. This function gives a higher weight to histogram bins indexed by bigger values of d. Then, by modifying the fuzzy histograms through this exponential function z^d, the invariant feature $\tilde{\mathbf{f}}^l(\mathcal{S}_g)$ can be redefined as follows:

$$\tilde{\mathbf{f}}^l(\mathcal{S}_g) = \left\langle z^d \, \mathcal{H}_{ld}^q(\mathcal{S}_g); q \in \{1, \ldots, Q_l\} \right\rangle. \tag{5.40}$$

Figure 5.8 resumes the proposed modifications described before. It shows the features $\tilde{\mathbf{f}}^l(\mathcal{S}_g)$ from Eq. (5.40) with $z = 10$ and compares

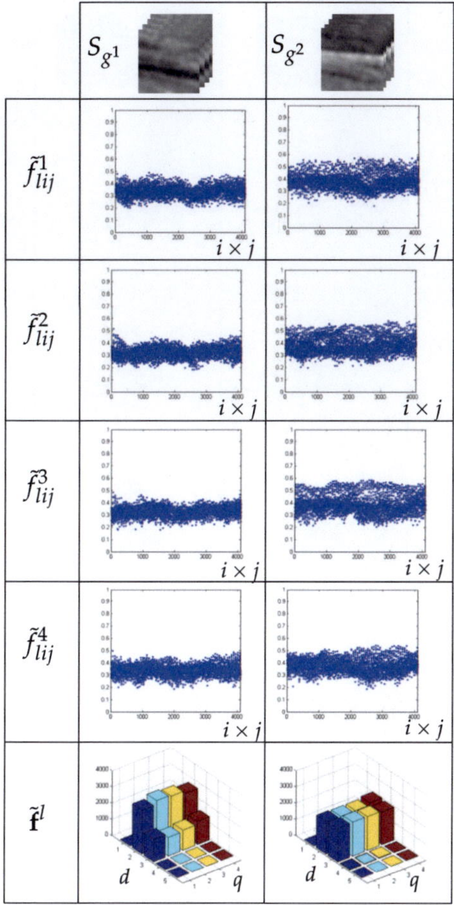

Figure 5.7: $\tilde{f}_{lij}^{q}(\mathcal{S}_g)$ obtained by $\mathbf{f}^l(\mathcal{S}_g)$ with $\mathbf{w}^l = (0, 3, 0°, 0°, 1, 90°)$ (Eq. (5.39)) for $\mathcal{S}_{g1}$ (intact surface) and $\mathcal{S}_{g2}$ (fissure), and the resulting features $\tilde{\mathbf{f}}^l(\mathcal{S}_g)$.

them with those obtained in Fig. 5.6 and Fig. 5.7. For $\mathcal{S}_{g2}$ in the second and third rows of Fig. 5.8, some relevant information about the defect (in this case, a fissure) is contained in the fourth bins of the histograms $\mathcal{H}_{ld}^{q}$ ($d = 4$). However, when no exponential function is used ($z = 1$, sec-

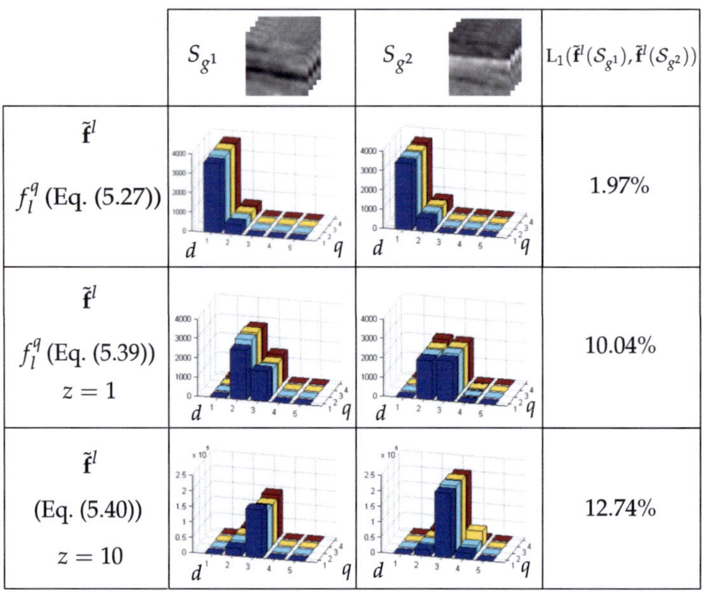

Figure 5.8: Comparison of the features $\tilde{\mathbf{f}}^l(\mathcal{S}_g)$ obtained for two series of images: $\mathcal{S}_{g1}$ (intact surface) and $\mathcal{S}_{g2}$ (fissure) using $\mathbf{w}^l = (0, 3, 0°, 0°, 1, 90°)$. 1st. row: Kernel function f^l (Eq. (5.15)) and $z = 1$. 2nd row: Kernel function f^l (Eq. (5.39)) and $z = 1$. 3 row: Kernel function f^l (Eq. (5.39)) and $z = 10$.

ond row), this information is almost imperceptible in comparison with other bin values. On the other hand, when $z = 10$ (third row), the fourth bins acquire more weight within the feature, which emphasizes the information concerning the defect. The fourth column of the figure shows the Manhattan distance $L_1(\tilde{\mathbf{f}}^l(\mathcal{S}_{g1}), \tilde{\mathbf{f}}^l(\mathcal{S}_{g2}))$ [DHS01] between the invariants obtained with the different approaches. This measure shows the separability improvement between invariant features when including the proposed modifications. As can be seen, the Manhattan distance increases in around 11% when the discussed modifications are applied. However, as the classification is performed by SVM, the real effect of the function z^d cannot be easily calculated and is going to be analyzed by means of an example in Section 7.3.

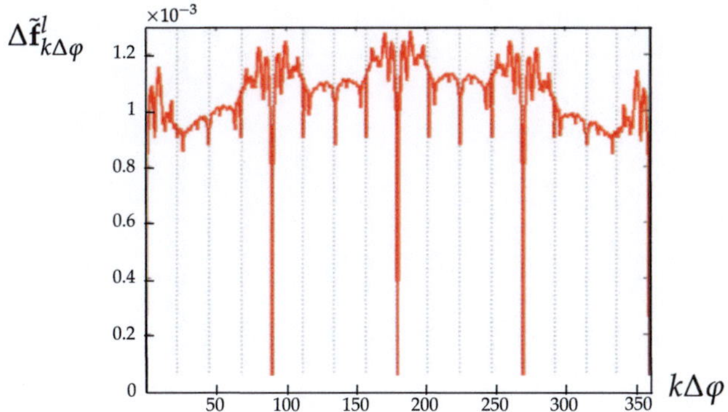

Figure 5.9: Divergence of the invariants $\tilde{\mathbf{f}}^l(\mathcal{S}_g)$ for different rotation angles $k\Delta\varphi$ and a series of images with $B=16$.

For the remainder of this thesis, the exponential weighing z^d is going to be considered as a part of the invariant feature $\tilde{\mathbf{f}}^l(\mathcal{S}_g)$.

5.3.4 Evaluating invariant properties of features

In this section, some tests are performed to measure the tolerance of the proposed invariant $\tilde{\mathbf{f}}^l(\mathcal{S}_g)$ against different illumination and kernel function parameters. The next two subsections deal with the analysis of invariance against rotation and translation, respectively.

Invariance against 2D rotation

In Section 5.2.1, the 2D rotation transformation was restated to consider series of images. In this case, the rotation involves a cyclic shift of intensity values through the third dimension given by b (Eq. (5.24)). This way, the maximum resolution for the rotation transformation is determined to $\Delta\varphi = \Delta\omega$ (Eq. (5.25)). That is, $\tilde{\mathbf{f}}^l(\mathcal{S}_g)$ can achieve invariance against rotation only when the object rotates in some angle $k\Delta\varphi$ that coincides with one of the used azimuth angles of the illuminant $b\Delta\omega$. However, as discussed previously, a system of invariants should be tolerant: It is expected that the values of $\tilde{\mathbf{f}}^l(\mathcal{S}_g)$ do not change significantly when the

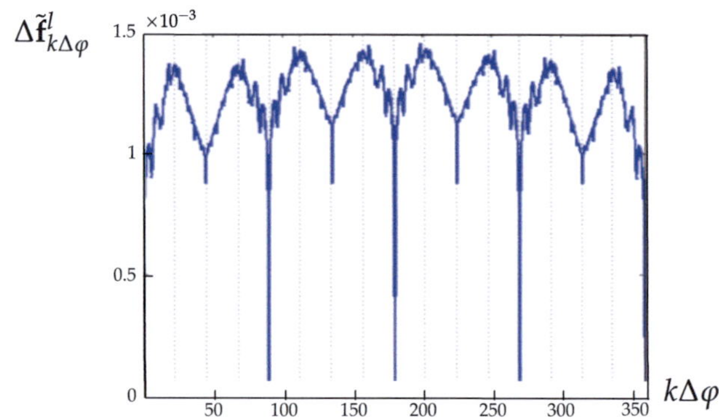

Figure 5.10: Divergence of the invariants $\tilde{\mathbf{f}}^l(\mathcal{S}_g)$ for different rotation angles $k\Delta\varphi$ and a series of images with $B=8$.

object rotates in some angle $k\Delta\varphi \neq b\Delta\omega$. On the contrary, they should remain as similar as possible in order not to affect the classification.

To test the tolerance of $\tilde{\mathbf{f}}^l(\mathcal{S}_g)$ to rotation angles $k\Delta\varphi \neq b\Delta\omega$, a series of images $\mathcal{S}_g$ is rotated from $k\Delta\varphi = 0°$ to $k\Delta\varphi = 360°$ considering $\Delta\varphi = 0.5°$. The resulting series of images after each rotation are denoted by $\mathcal{S}_{g(k\Delta\varphi)}$. For each rotated series $\mathcal{S}_{g(k\Delta\varphi)}$, the proposed invariant feature is calculated and denoted by $\tilde{\mathbf{f}}^l_{k\Delta\varphi} := \tilde{\mathbf{f}}^l(\mathcal{S}_{g(k\Delta\varphi)})$. Ideally, the invariants $\tilde{\mathbf{f}}^l_{k\Delta\varphi}$ should result identical each time that $k\Delta\varphi = b\Delta\omega$, for some $b \in \{0,\ldots,B-1\}$ and $k \in \{0,\ldots,K-1\}$, and remain very similar for the other values of $k\Delta\varphi$. However, as the rotation transformation is simulated from one series of images, additional divergences are expected because of interpolation errors. Taking the invariant obtained from the original series of images $\tilde{\mathbf{f}}^l_{0°}$ as a reference, a measure for the tolerance of the feature against each rotation angle $k\Delta\varphi$ can be defined as follows:

$$\Delta\tilde{\mathbf{f}}^l_{k\Delta\varphi} = \frac{1}{Q_l D} \sum_{q=1}^{Q_l} \sum_{d=1}^{D} |\tilde{\mathbf{f}}^l_{0°} - \tilde{\mathbf{f}}^l_{k\Delta\varphi}|. \tag{5.41}$$

Equation (5.41) sums up the difference between elements of the invariants $\tilde{\mathbf{f}}^l_{0°}$ and $\tilde{\mathbf{f}}^l_{k\Delta\varphi}$. Each element of these invariants is indexed by the

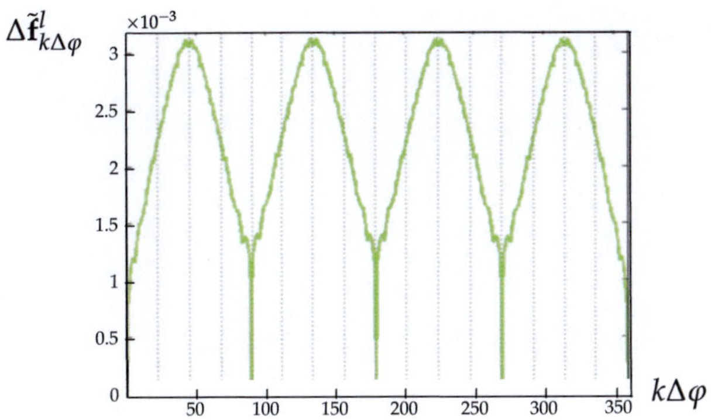

Figure 5.11: Divergence of the invariants $\tilde{\mathbf{f}}^l(\mathcal{S}_g)$ for different rotation angles $k\Delta\varphi$ and a series of images with $B=4$.

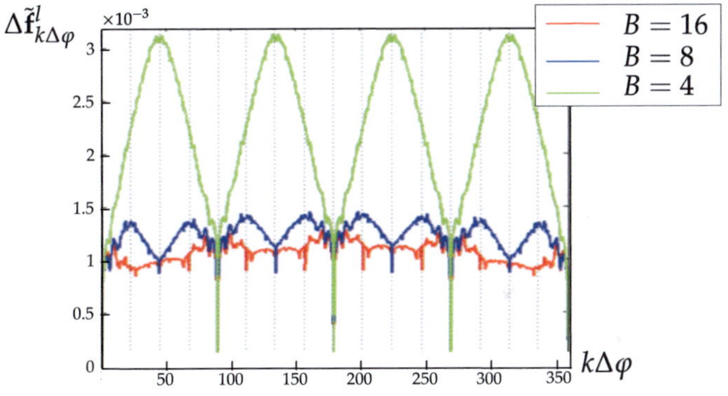

Figure 5.12: Divergence of the invariants $\tilde{\mathbf{f}}^l(\mathcal{S}_g)$ for different rotation angles $k\Delta\varphi$ for series of images with $B \in \{4, 8, 16\}$.

parameters q and d (Eq. (5.40)). The divergence $\Delta\tilde{\mathbf{f}}^l_{k\Delta\varphi}$ is normalized and presents values in $[0, 1]$. Thus, $\Delta\tilde{\mathbf{f}}^l_{k\Delta\varphi} = 0$ indicates that the invariants are identical and $\Delta\tilde{\mathbf{f}}^l_{k\Delta\varphi} = 1$ corresponds to the maximal difference between them. Figure 5.9 shows the divergence for a series of images

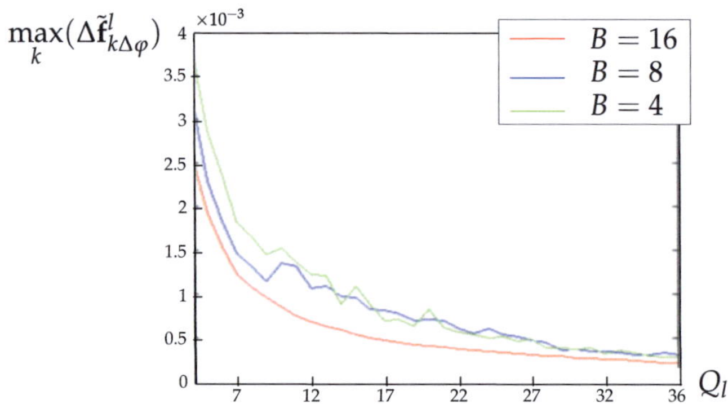

Figure 5.13: Maximal divergence of the invariants for a complete rotation considering series of images with $B \in \{4, 8, 16\}$ and different values of Q_l.

of a fissure with $B = 16$. A fissure was selected to perform this test because its shape makes it specially sensitive to rotation. The invariants were calculated with $\mathbf{w}^l = (0, 3, 0°, 0°, 1, 90°)$ and $z = 1$. The maximal divergence value obtained in this example for a complete rotation was 0.001282, i.e., the proposed feature presents a maximal discrepancy of 0.12% for the rotation transformation. As $B = 16$ ($\Delta\omega = 22.5°$), the invariants should be more robust for rotation angles multiple of 22.5°. This latter can be observed in Fig. 5.9 as $\Delta\tilde{\mathbf{f}}^l_{k\,22.5°}$ tends to zero. $\Delta\tilde{\mathbf{f}}^l_{k\Delta\varphi}$ reaches zero at $k\Delta\varphi \in \{0°, 90°, 180°, 270°\}$, because these angles do not present interpolation errors for the simulated rotation. Figure 5.10 shows $\Delta\tilde{\mathbf{f}}^l_{k\Delta\varphi}$ for $B = 8$ ($\Delta\omega = 45°$). This time, the maximal divergence is 0.001452 and $\Delta\tilde{\mathbf{f}}^l_{k\Delta\varphi}$ tends to zero each 45°. Further, Figure 5.11 shows $\Delta\tilde{\mathbf{f}}^l_{k\Delta\varphi}$ for $B = 4$ with a maximal divergence of 0.003122. Figure 5.12 reproduces all results together for $B \in \{4, 8, 16\}$. Clearly, from Fig. 5.12, the invariance against rotation increases with the number of images in the series. This is because more images bring along more information about the object. On the other hand, larger series of images are associated with more measurement and computational cost. The selected kernel function parameters in $\mathbf{w}^l$ also affect the tolerance of invariants against rotation. Specially, $\Delta\rho_l$ has a big influence on the tolerance be-

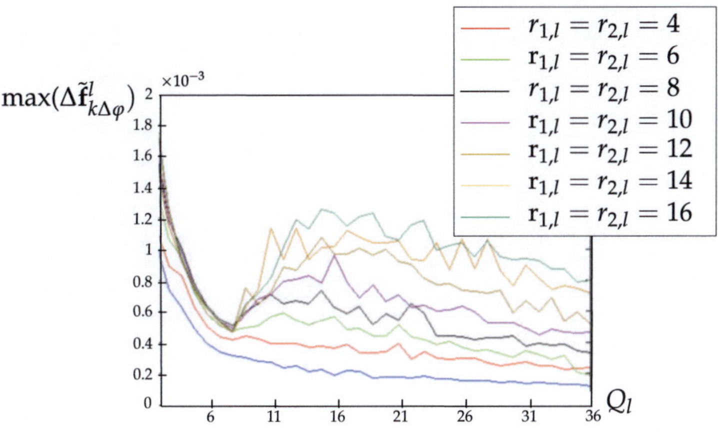

Figure 5.14: Maximal divergence of the invariants for a complete rotation considering a series of images with $B = 8$ and different values of Q_l, $r_{1,l}$ and $r_{2,l}$. The radii $r_{1,l}$ and $r_{2,l}$ are considered to be equal.

cause it determines the sample density of the circular neighborhoods given by $r_{1,l}$ and $r_{2,l}$ and therefore the amount of information collected by the kernel function. To illustrate this, the maximum divergence $\max_k(\Delta \tilde{\mathbf{f}}^l_{k\Delta\varphi})$ is calculated for different values of $\Delta\rho_l$, i.e., different values of Q_l. For the same series of images used before (which reproduces a fissure), Fig. 5.13 plots $\max_k(\Delta \tilde{\mathbf{f}}^l_{k\Delta\varphi})$ considering $Q_l \in \{2, \ldots, 36\}$ and $B \in \{4, 8, 16\}$. It can be observed that the invariance against rotation augments for bigger values of Q_l and B. However, the difference in the divergence given by B tends to zero as Q_l increases. Of course, big values of Q_l are associated to more computational cost.

Other elements of $\mathbf{w}^l$ that influence the feature tolerance are the radii $r_{1,l}$ and $r_{2,l}$. Longer radii lead to larger neighborhoods, which increases the effect of inhomogeneous wood texture on results. Figure 5.14 plots $\max_k(\Delta \tilde{\mathbf{f}}^l_{k\Delta\varphi})$ for different values of Q_l, $r_{1,l}$ and $r_{2,l}$. As in the previous figures, the series of images shows a fissure, but this time B is fixed to 8. It can be concluded that the divergence tends to increase for large radii. Furthermore, for given values of $r_{1,l}$ and $r_{2,l}$, the divergence tends to

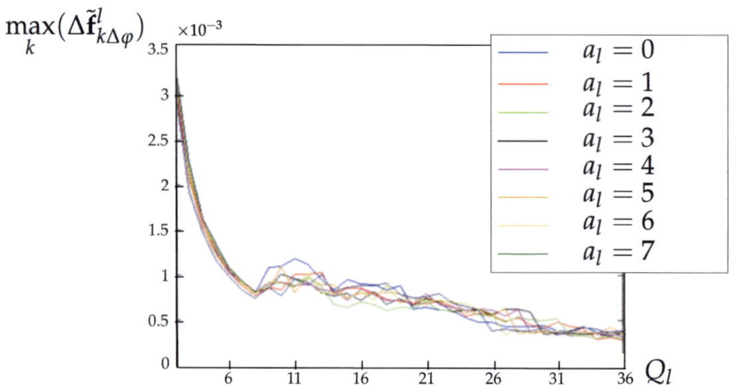

Figure 5.15: Maximal divergence of the invariants for a complete rotation considering a series of images with $B=8$ and different values of Q_l and a_l.

remain relatively constant for $Q_l \geq 10$. The remaining elements in $\mathbf{w}^l$, the angles α_l and β_l and a_l, have almost no influence on the invariant tolerance as showed in Fig. 5.15 and Fig. 5.16 (for the same series of images and $B = 8$).

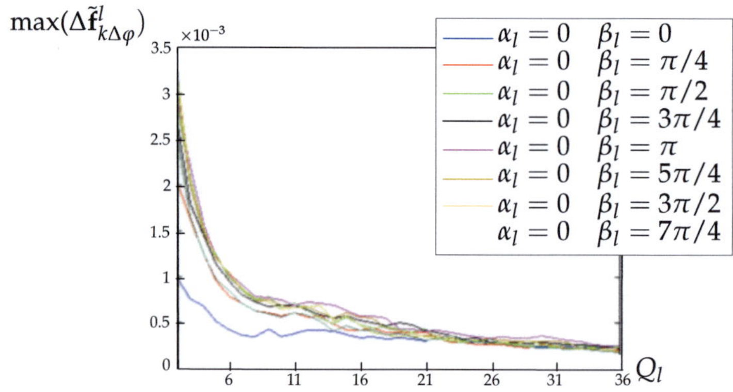

Figure 5.16: Maximal divergence of the invariants for a complete rotation considering a series of images with $B=8$ and different values of Q_l and β_l.

As shown in this section, the developed method achieves invariants $\tilde{\mathbf{f}}^l(\mathcal{S}_g)$ that are tolerant against the 2D rotation transformation. Furthermore, it can be concluded that the optimal parameters to reach robust invariants are given by $B \geq 8$, $Q_l \geq 8$ and $r_{1,l}, r_{2,l} \leq 6$.

Invariance against 2D translation

In this subsection, a translation is simulated on a series of images showing a bubble. The parameters i and j are varied from 0 to $M-1$ and from 0 to $N-1$ respectively. The resulting series of images after each translation is denoted by $\mathcal{S}_{g(ij)}$, whereas its corresponding invariant is called $\tilde{\mathbf{f}}^l_{ij} := \tilde{\mathbf{f}}^l(\mathcal{S}_{g(ij)})$. The divergence of invariants is calculated as follows:

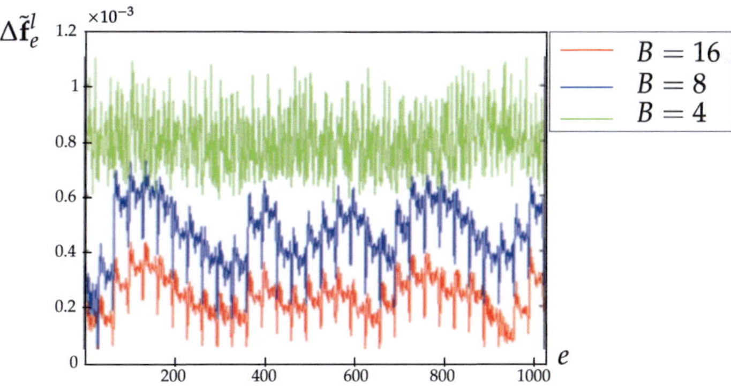

Figure 5.17: Divergence of the invariants for the translation transformation considering a series of images with $B \in \{4, 8, 16\}$.

$$\Delta \tilde{\mathbf{f}}^l_{ij} = \frac{1}{Q_l D} \sum_{q=1}^{Q_l} \sum_{d=1}^{D} |\tilde{\mathbf{f}}^l_{00} - \tilde{\mathbf{f}}^l_{ij}|. \tag{5.42}$$

To simplify plots of $\Delta \tilde{\mathbf{f}}^l_{ij}$, the translation transformation in both directions i and j can be linearly indexed by:

$$e = \sum_{i=0}^{M-1} \sum_{j=0}^{N-1} i + jM,$$

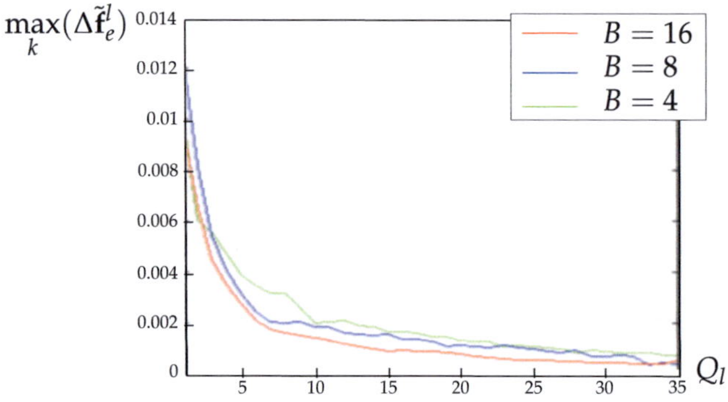

Figure 5.18: Maximal divergence of the invariants for the translation transformation considering series of images with $B \in \{4, 8, 16\}$ and different values of Q_l.

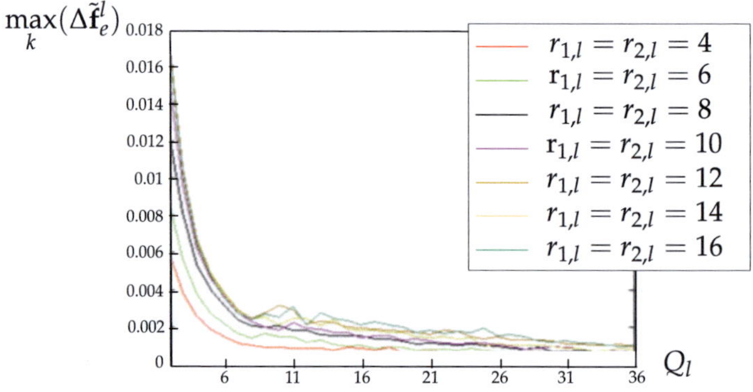

Figure 5.19: Maximal divergence of the invariants for the translation transformation considering a series of eight images, different values of Q_l, $r_{1,l}$ and $r_{2,l}$. The radii $r_{1,l}$ and $r_{2,l}$ are considered to be equal.

where $e \in \{0, \ldots, M \times N - 1\}$. As a consequence, Eq. (5.42) can be reshaped in the following way:

$$\Delta \tilde{\mathbf{f}}_e^l = \frac{1}{Q_l D} \sum_{q=1}^{Q_l} \sum_{d=1}^{D} |\tilde{\mathbf{f}}_0^l - \tilde{\mathbf{f}}_e^l|.$$

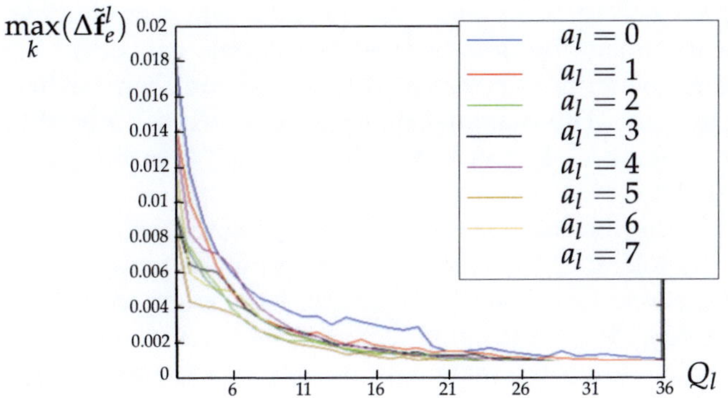

Figure 5.20: Maximal divergence of the invariants for the translation transformation considering a series of eight images, different values of Q_l and a_l.

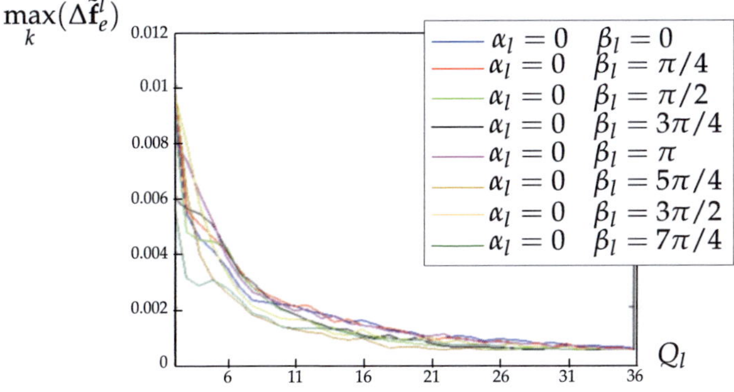

Figure 5.21: Maximal divergence of the invariants for the translation transformation considering a series of eight images, different values of Q_l and β_l.

Figure 5.17 plots the divergence for a series of images showing a bubble, considering $B = 16$, $B = 8$ and $B = 4$ (as in the case of rotation $\mathbf{w}^l = (0, 3, 0°, 0°, 1, 90°)$ and $z = 1$). For the presented plots, the translation has not been considered to be cyclical, which produces border effects (i.e., bigger divergence values are obtained when the bubble reaches some of the image borders because the bubble gets cut). In general, the feature tolerance is better for the translation than for the rotation transformation. The maximal divergence for $B = 16$ is of 0.0004316 ($\approx 0.04\%$), for $B = 8$ is of 0.00072522 ($\approx 0.07\%$) and for $B = 4$ is of 0.0011 (0.11%).

Considering the same series of images, Fig. 5.18 shows the relation among the tolerance against translation and the parameter Q_l of the kernel function. As in the case of rotation, the tolerance arises with the parameter Q_l.

Figure 5.19 shows the maximal divergence of the invariants with respect to the parameters $r_{1,l}$ and $r_{2,l}$ of the kernel function. Here again, the tolerance decreases with increasing radii. However, in the case of translation, the differences between the values of $\max_k(\Delta \tilde{\mathbf{f}}_e^l)$ for different radii are smaller than in the case of rotation.

Figures 5.20 and 5.21 show the maximal divergence for translation considering different values of the kernel function parameters a_l and β_l. As for rotation, these parameters have no influence on the divergence values.

6 Classification

This chapter is concerned with the classification of the system of invariants $\tilde{\mathcal{F}}_g := \tilde{\mathcal{F}}(\mathcal{S}_g)$ of Eq. (5.38). As already discussed, the classification procedure consists in assigning a class $c \in \mathcal{C}$ to each $\tilde{\mathcal{F}}_g$:

$$c_f : \mathcal{F}^L \longrightarrow \mathcal{C}; \quad \tilde{\mathcal{F}}_g \mapsto c.$$

This class assignment is performed by a classifier, which is defined as anything that takes a feature set as an input and delivers a class label as an output [FP03]. In this work, the selected classifier is a Support Vector Machine (SVM). The foundations of SVM have been developed in the late seventies [Vap79], but SVM began to receive a bigger attention in the nineties. The formulation of SVM embodies the Structural Risk Minimization (SRM) principle, which has been shown to be superior to the traditional Empirical Risk Minimization (ERM) principle used by conventional neuronal networks. A short introduction to SVM theory is presented in appendix B.

6.1 Training process

The classification of the inspected surface is performed locally. This means that the surface is divided into regions, which are then analyzed to decide whether they contain certain classes of defects or not. As a consequence, the series of images $\mathcal{S}_g$ must also be separated into regions. These regions are called windows and are denoted by $\mathcal{W}_g \subset \mathcal{S}_g$. The window size is given by $M'' \times N'' \times B$ pixels, where M'' and N'' are divisors of M and N respectively. These windows are successively extracted from $\mathcal{S}_g$ with a vertical and horizontal overlap given by $\vartheta < 100\%$. Then, from a series of images $\mathcal{S}_g$, a set of windows $\{\mathcal{W}_g^w \mid w \in \{1, \ldots, W\}\}$ can be extracted, where:

$$W = \left(\left(1 - \frac{\vartheta}{100}\right)^{-1} \cdot \frac{M}{M''} - 1 \right) \left(\left(1 - \frac{\vartheta}{100}\right)^{-1} \cdot \frac{N}{N''} - 1 \right).$$

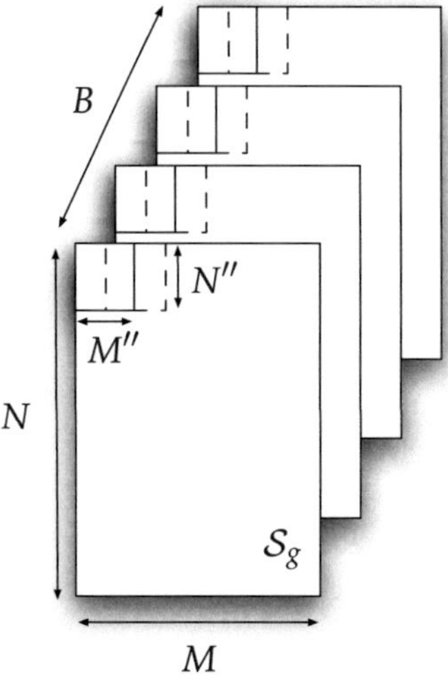

Figure 6.1: Extraction of windows $\mathcal{W}_g^w$ from a series of images $\mathcal{S}_g$, with $B = 4$ and $\vartheta = 50\%$.

Figure 6.1 illustrates the extraction of the windows $\mathcal{W}_g^w$ from a series of images with $B = 4$ and $\vartheta = 50\%$. For each window $\mathcal{W}_g^w$ (not for the whole series of images $\mathcal{S}_g$), the system of invariants is calculated and classified. In order to consider this local analysis in the notation, the system of invariants is denoted by $\tilde{\mathcal{F}}_{gw} := \tilde{\mathcal{F}}(\mathcal{W}_g^w)$ with $\mathcal{W}_g^w \subset \mathcal{S}_g$.

A representative set of series of images $\mathcal{L} = \{\mathcal{S}_{g^s} | s \in \{1, \ldots, S\}\}$, with $S \in \mathbb{N}$, is used to train the SVM. All considered classes in $\mathcal{C}$ must be well represented in $\mathcal{L}$ for a successful training. As the SVM is a supervised classifier, it is necessary to manually indicate whether and where there is a defect in $\mathcal{L}$ and to which class it belongs. Consequently, the class $c_{gw} \in \mathcal{C}$ of each window $\mathcal{W}_g^w$ must be known for each series of images $\mathcal{S}_g \in \mathcal{L}$. That is, for each series of images in $\mathcal{L}$, the windows $\mathcal{W}_{gw}$ are extracted, then the systems of invariants $\tilde{\mathcal{F}}_{gw}$ are calculated and finally the pairs $(\tilde{\mathcal{F}}_{gw}, c_{gw})$ are used to train.

In order to assign a class c_{gw} to each system of invariants $\tilde{\mathcal{F}}_{gw}$, a mask $\breve{g}_{mn}$ must be built from each series of images $\mathcal{S}_g \in \mathcal{L}$. The mask $\breve{g}_{mn}$ is an $M \times N$ image, whose gray-level values are between 0 and $|\mathcal{C}| - 1$. For generating this mask, an expert analyzes the series of images and covers each existing defect with windows of size $M'' \times N''$ indicating the corresponding class. Different defects or even different parts of defects can be marked in different images g_{mnb} of $\mathcal{S}_g$. The expert has only to search the image of the series in which defects are more visible and therefore can be more precisely delimited. Once all defects in a series of images have been covered and their classes have been determined, the mask $\breve{g}_{mn}$ is generated as follows:

$$\breve{g}_{\hat{m}\hat{n}} = \begin{cases} 0 & g_{\hat{m}\hat{n}b} \text{ was not tagged as defect for any } b, \\ c & g_{\hat{m}\hat{n}b} \text{ was tagged as defect class } c \text{ for some } b. \end{cases} \qquad (6.1)$$

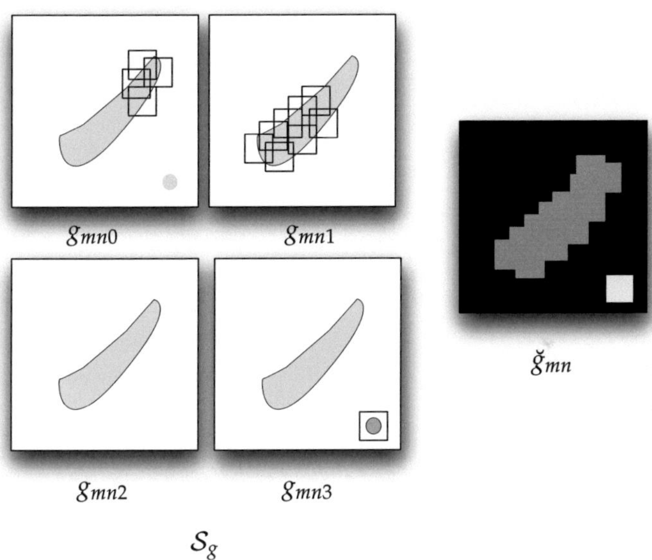

Figure 6.2: Generating a mask from a series of images $\mathcal{S}_g$ with $B = 4$. The series of images shows two different defects. The biggest defect has been marked off by the expert in the first and second image, while the smallest one was found in the last image. The black background in the mask represents the class zero, i.e., no defect, whereas the other two gray levels denote different defect classes.

Figure 6.2 illustrates the process of creating a mask for a four-image series reproducing two defects.

Once a mask $\breve{g}_{mn}$ is generated for each series of images in $\mathcal{L}$, the training process can be started. For each window $\mathcal{W}_g^w$ extracted from $\mathcal{S}_g$, a window $\breve{m}_g^w$ is also extracted from the corresponding mask $\breve{g}_{mn}$ so as to assign a class c_{gw} to each system of invariants $\tilde{\mathcal{F}}_{gw}$. Finally, the class c_{gw}, which is given by the most frequent gray level in $\breve{m}_g^w$, is the one assigned to $\tilde{\mathcal{F}}_{gw}$. Figure 6.3 illustrates the window extraction from a series of images and its corresponding mask.

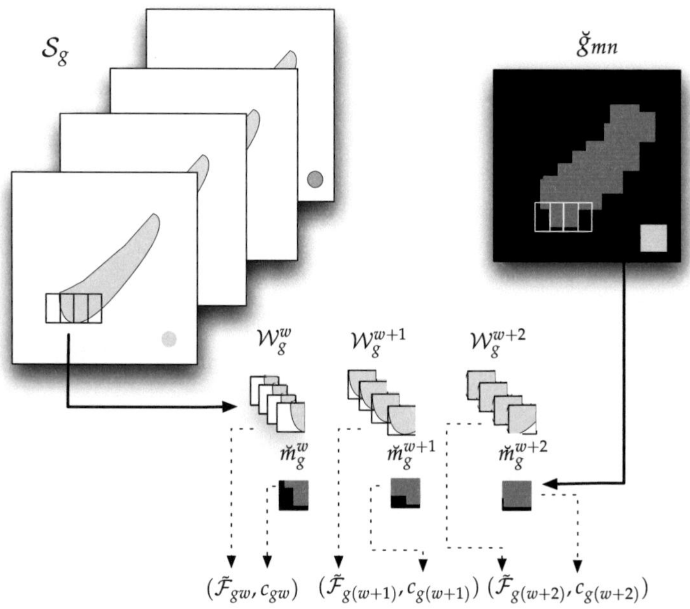

Figure 6.3: Extraction of windows from a series of images and its corresponding mask.

Once the complete training data set $\{(\tilde{\mathcal{F}}_{g^s w}, c_{g^s w})\}$ with $s \in \{1, \ldots, S\}$ and $w \in \{1, \ldots, W\}$ is obtained, the SVM can be trained. The SVM implementation used in this thesis is the LIBSVM. For solving multi-class problems, like the one presented in this thesis, the LIBSVM uses the "one-agains-one" approach, see [CL01]. A previous important step to training is scaling features [HCL08]. In this work, the scaling must be performed considering that features are not real numbers but matrices of real numbers. However, the LIBSVM interprets each value in $\tilde{\mathbf{f}}^l(\mathcal{S}_g)$ as an individual feature. The scaling process in LIBSVM is thus performed by grouping features in sets denoted by $\{\tilde{f}^q_{lij}(\mathcal{S}_g)|l \in \{1,\dots,L\}\}$ according to each value of i, j and q. This leads to a false interpretation of the feature and therefore to false classification results. For this reason, the scale process of the LIBSVM must be modified to group the features in sets given by $\{\tilde{\mathbf{f}}^l(\mathcal{S}_g)|l \in \{1,\dots,L\}\}$.

Summarizing, a successful training depends on many factors: the quality of the series of images $\mathcal{S}_g$, a good selection of the training set $\mathcal{L}$, a correct generation of masks, a suitable choice of window sizes and of kernel function parameters, and an adequate configuration of the SVM.

6.2 Testing process

In order to test a series of images $\mathcal{S}_g$, windows $\mathcal{W}^w_g$ must also be extracted. Obviously, in the case of testing, it is not necessary to generate a mask. The testing can be summarized as follows: Given an unknown series of images $\mathcal{S}_g$, windows $\mathcal{W}^w_g$ are extracted in the same way as for the training. Then, for each window, the system of invariants $\tilde{\mathcal{F}}_{gw}$ is calculated and passed on to the SVM, which answers with a predicted class $\hat{c}_{gw}$ (i.e., $\hat{c}_{gw}$ is the result of the classification). Further, if the mask of $\mathcal{S}_g$ is available, comparing the predicted class $\hat{c}_{gw}$ with the class c_{gw} given by the mask allows evaluating the classification result.

6.2.1 Confusion matrix and classification rates

Results can be evaluated by comparing the predicted class $\hat{c}_{gw}$ of each window $\mathcal{W}^w_g$ from the tested $\mathcal{S}_g$ with the class c_{gw} of the corresponding mask window $\check{m}^w_g$. The resulting pairs $(c_{gw}, \hat{c}_{gw})$ are used to construct a confusion matrix CM. CM is a $|\mathcal{C}| \times |\mathcal{C}|$ matrix, whose elements are denoted by $(c, \hat{c})$ where c and $\hat{c}$ belong to $\mathcal{C}$. Each element $(c, \hat{c})$ of CM is a counter, which is incremented each time a window of class c is classified as $\hat{c}$. Therefore, the confusion matrix allows analyzing the classification results per class. It delivers information such as which defects are more frequently correctly classified, which defects are more frequently confused with other ones and which are more difficult to detect. Figure 6.4 schematizes a confusion matrix considering 7 classes, $c = 0$ describes always the no-defect class. The columns in CM represent the correct classes given by masks, while the rows stand for predicted classes from the SVM. All values are normalized per class according to the mask information, i.e., each column of CM sums 100%. To analyze the whole classification performance, the CM can be divided into five regions: false negatives, true negatives, false positives, true positives/true classifications and true positives/true detections. The first element of CM gives the percentage of true negatives tn, i.e., the percentage of windows without defects that are correctly classified as no-defect windows:

$$tn = CM(c, \hat{c}) \quad \text{for } c = \hat{c} = 0. \tag{6.2}$$

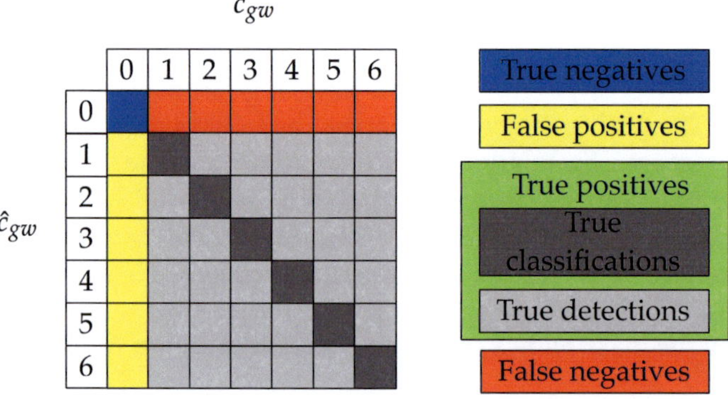

Figure 6.4: Confusion matrix for $|\mathcal{C}| = 7$.

The other elements of CM's first row show the false negatives for each class, i.e., the percentage of each defect class that was not detected at all. The whole false negative rate fn is given by:

$$fn = \frac{1}{|\mathcal{C}| - 1} \sum_{c=1}^{|\mathcal{C}|-1} CM(c, \hat{c} = 0). \tag{6.3}$$

The first column of CM, without considering the first element, gives the false positives per class. The whole false positive rate fp is given by:

$$fp = \frac{1}{|\mathcal{C}| - 1} \sum_{\hat{c}=1}^{|\mathcal{C}|-1} CM(c=0, \hat{c}). \tag{6.4}$$

The remaining elements of the matrix represent the true positives tp. This latter can be divided into true classification given by values on the diagonal and true detection given by all other elements. The true classification tc denotes those defects that were detected and correctly classified, while the true detection td involves those defects that were detected but wrongly classified:

$$tc = \frac{1}{|\mathcal{C}| - 1} \sum_{c=1}^{|\mathcal{C}|-1} \sum_{\hat{c}=1}^{|\mathcal{C}|-1} CM(c, \hat{c}) \quad \forall\, \hat{c} = c, \tag{6.5}$$

$$td = \frac{1}{|\mathcal{C}| - 1} \sum_{c=1}^{|\mathcal{C}|-1} \sum_{\hat{c}=1}^{|\mathcal{C}|-1} CM(c, \hat{c}) \quad \forall\, \hat{c} \neq c, \tag{6.6}$$

$$tp = tc + td. \tag{6.7}$$

The confusion matrix CM and its derived performance measures (tn, fn, fp, etc.) are going to be applied in the next chapter to evaluate the obtained results.

7 Results

The proposed method is evaluated in the classification of six different defects $|\mathcal{C}| = 7$ (i.e., 6 defects plus the no-defect class). Figure 7.1 shows examples of each defect class and introduces a color code that helps interpret masks $\breve{g}_{mn}$ and classification results. For the purpose of this chapter, a set $\mathcal{L}$ of 85 series of images (i.e., $S = 85$) is considered. The first image of each $\mathcal{S}_g \in \mathcal{L}$ and the corresponding masks $\breve{g}_{mn}$ are presented in Appendix C. Almost all series of images in $\mathcal{L}$ show wood surfaces with transparent varnish, so that wood texture is visible in most cases. Furthermore, ten different substrates are included in $\mathcal{L}$, which are all shown in Fig. 7.2. The evaluation of results is performed using the Leave One Out (LOO) method. The LOO method consists in training the SVM with all series of images in $\mathcal{L}$ except for one. The testing is then carried out on the series of images that was not considered in the training. This procedure is repeated for all series in $\mathcal{L}$. As a consequence, the SVM is trained S times. After each training, the W windows of the excluded series of images must be tested, which results in $S \cdot W$ test repetitions for evaluating the whole set $\mathcal{L}$.

7.1 Resolution and window sizes

The dimensions ($M'' \times N''$ pixels) of windows $\mathcal{W}_g^w$ have a big influence on the classification results and must be carefully selected according to the resolution of images. In order to find the most convenient configuraten of resolution and window size, a subset $\mathcal{L}' \subset \mathcal{L}$, in which all considered classes are represented, is taken into account for $B = 8$. Using different resolutions for the series of images $\mathcal{S}_g \in \mathcal{L}'$, three versions of $\mathcal{L}'$ are generated for 32, 16 and 8 pixels/mm: $\mathcal{L}'_{32}$, $\mathcal{L}'_{16}$ and $\mathcal{L}'_8$. Further, for each of these, three sets of masks are constructed for windows of 16×16, 32×32 and 64×64 pixels. Figure 7.3 schematizes the relation between different resolutions and window sizes considering that always the same surface of 1.6×1.6 cm^2 is imaged. In this figure, rows show the different window sizes, while columns represent the consid-

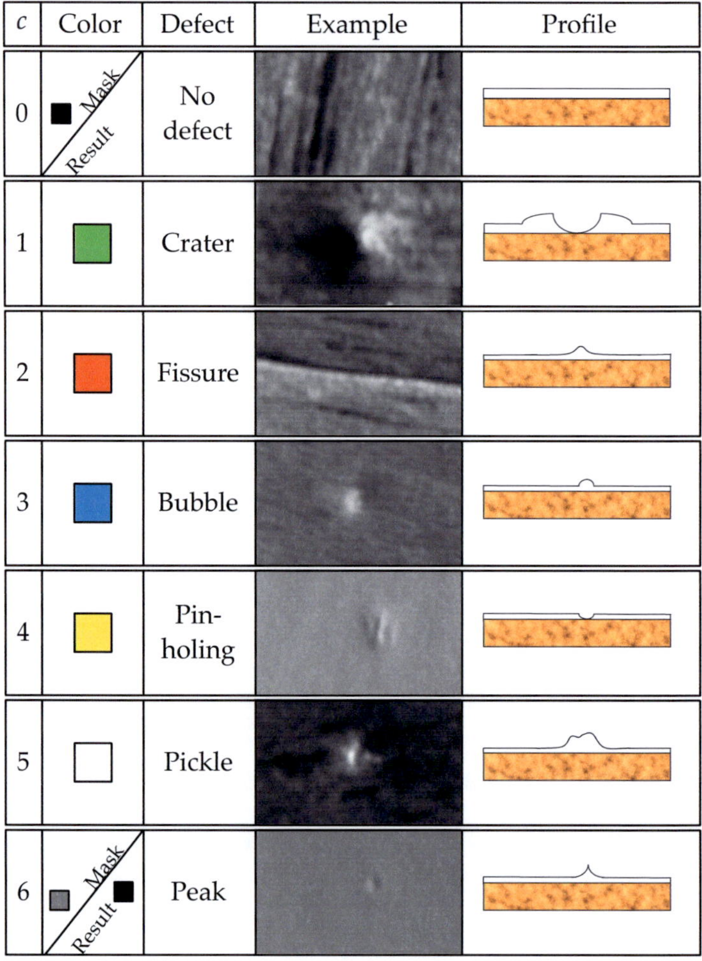

Figure 7.1: Different classes considered in $\mathcal{C}$. The color code helps interpret masks and classification results. The fissure shows a linear symmetry, while all other defects have a circular symmetry [PAP08]. For the no-defect and peak classes, different colors are used to generate masks and classification results.

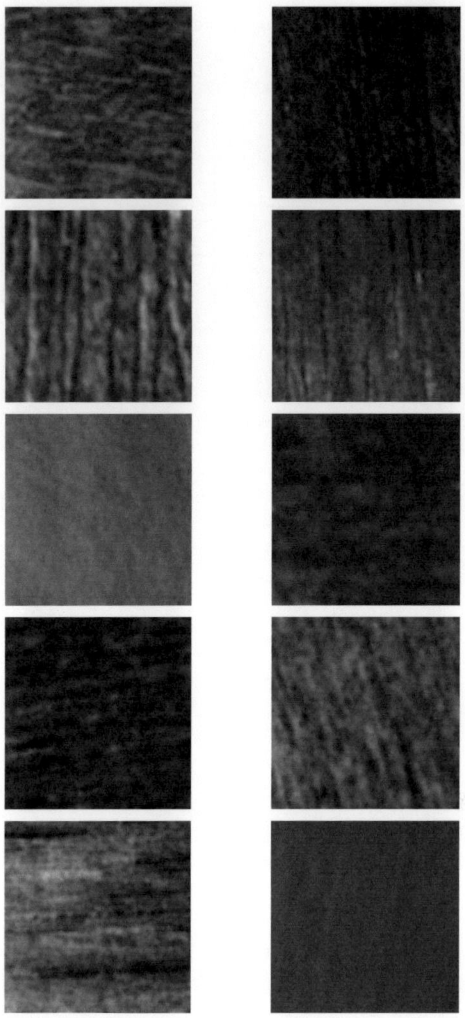

Figure 7.2: Different wood textures considered in $\mathcal{L}$.

ered image resolutions. For 32 pixels/mm (first column), the images $g_{mnb} \in \mathcal{S}_g$ have a size of 512×512 ($M \times N$) pixels, while, for the reso-

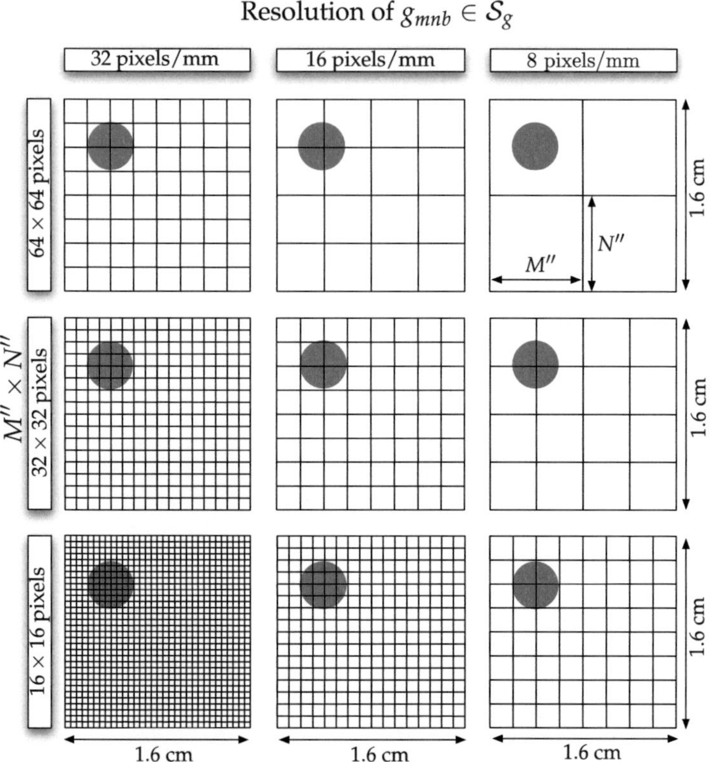

Figure 7.3: Relation between different image resolutions and window sizes. Columns represent image resolutions of 32, 16 and 8 pixels/mm, whereas rows represent window sizes of 64 × 64, 32 × 32 and 16 × 16 pixels.

lutions of 16 and 8 pixels/mm (second and third columns), the images have respectively 256 × 256 and 128 × 128 pixels. For an overlap factor $\vartheta = 0\%$ and a window of 64 × 64 pixels, 64 windows can be extracted if a resolution of 32 pixels/mm is used. On the other hand, 16 windows can be obtained for a resolution of 16 pixels/mm and only 4 windows for a resolution of 8 pixels/mm. Now, to find out which combination of resolution and window size is the most appropriate to classify varnish defects, a system of invariants $\tilde{\mathcal{F}}_g$ is created. In order not to opti-

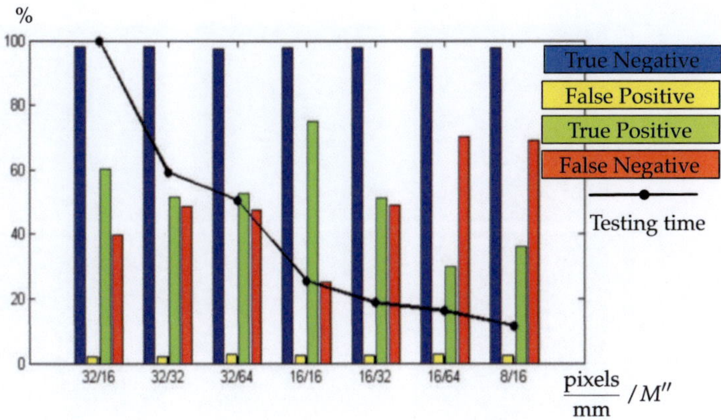

Figure 7.4: Relative classification rates tn, tp, fn and fp for different resolutions and window sizes. Testing time is normalized with regard to its respective maximum.

mize $\tilde{\mathcal{F}}_g$ to any resolution or window size, the invariants $\tilde{\mathbf{f}}^l(\mathcal{S}_g) \in \tilde{\mathcal{F}}_g$ were randomly generated. Applying the LOO method, confusion matrices CM are constructed for $\mathcal{L}'_{32}$, $\mathcal{L}'_{16}$ and $\mathcal{L}'_8$ considering each time the three given window sizes. From each resulting CM, relative classification rates tp, tn, fn and fp are calculated (based on Eqs. (6.2) to (6.7)) and plotted in Fig. 7.4. The results for $\mathcal{L}'_8$ and window sizes of 32×32 and 64×64 pixels are omitted from Fig. 7.4, because windows are much bigger than defects under these conditions. So, it is neither possible to generate masks nor to classify defects. Figure 7.4 shows that using 16 pixels/mm with windows of 64×64 pixels gives the highest rates of false positives and negatives. All combinations of resolution and window size describe almost the same true negative rate. The highest true positive rate is given at the 16 pixels/mm resolution with windows of 16×16 pixels. The second best combination is given by 32 pixels/mm and windows of 16×16 pixels, which presents $\approx 15\%$ less true positives but also a little less false positives. Additionally, the combination of a high image resolution with a small window size increases the computational cost of the classification. Figure 7.4 shows the normalized testing time, which involves the time used to calculate and scale the systems

95

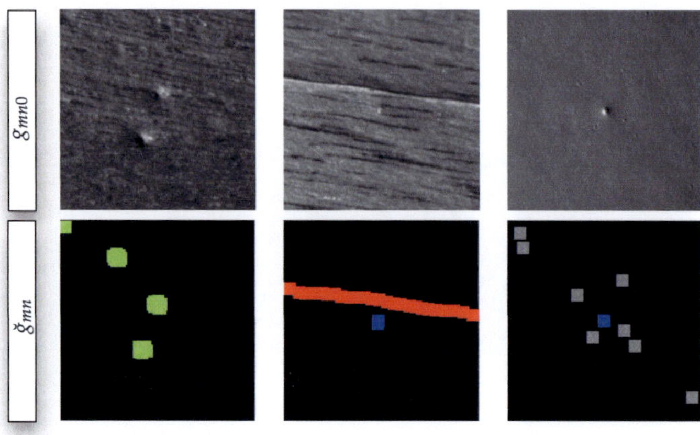

Figure 7.5: First images g_{mn0} of three series $S_g \in \mathcal{L}'$ and the respective masks. The images show craters (first column), a fissure and a bubble (second column) and a bubble and several peaks (third column). Note that the masks are constructed using the color code of Fig. 7.1.

of invariants, to perform the testing and to predict the corresponding class. In terms of classification rates, image resolutions of 32 and 16 pixels/mm with windows of 16×16 pixels lead to better results. However, using 16 pixels/mm requires $\approx 75\%$ less testing time. As a consequence, a more efficient classification is achieved with a resolution of 16 pixels/mm and a window size of 16×16 pixels.

Three different series of images $\mathcal{S}_g$ of $\mathcal{L}'$ have been selected to illustrate the influence of image resolution and window size on the classification with respect to different defect sizes. The first image of these series and the respective masks are shown in Fig. 7.5. The sizes of the imaged defects are very different: from big craters and midsize fissures to small bubbles and very small peaks. Figures 7.6, 7.7 and 7.8 show the classification results of the selected series of images for the different considered resolutions and window sizes. As stated before, the best results are obtained using resolutions of 32 and 16 pixels/mm and a window size of 16×16 pixels.

7.2 Number B of images per series

To select the combination of image resolution and window size, the classification performance has been measured on series of eight images ($B = 8$). Now, fixing the resolution to 16 pixels/mm and the window size to 16×16 pixels, the influence of B on the classification is going to be analyzed. From the previous subset $\mathcal{L}'_{16}$, three versions are going to be considered: $\mathcal{L}'_{16/16}$, where each series of images has $B = 16$, $\mathcal{L}'_{16/8}$ with $B = 8$ and $\mathcal{L}'_{16/4}$ with $B = 4$. Reducing B saves up time not only on the calculation of the rotation invariants $\tilde{\mathbf{f}}^q_{lij}(\mathcal{S}_g)$ (in the case that $\Delta\varphi$ is directly determined by $\Delta\omega$), but also during the image acquisition.

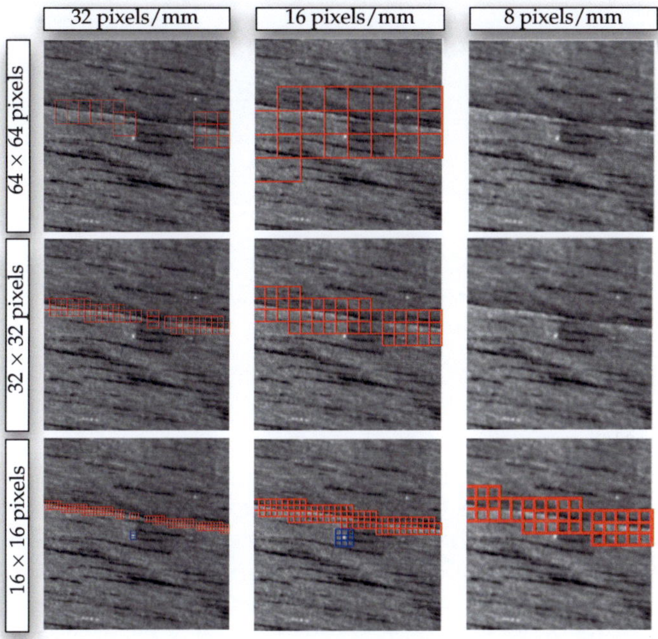

Figure 7.6: Classification results for a fissure and a bubble using different resolutions and window sizes. The color code of Fig. 7.1 was used.

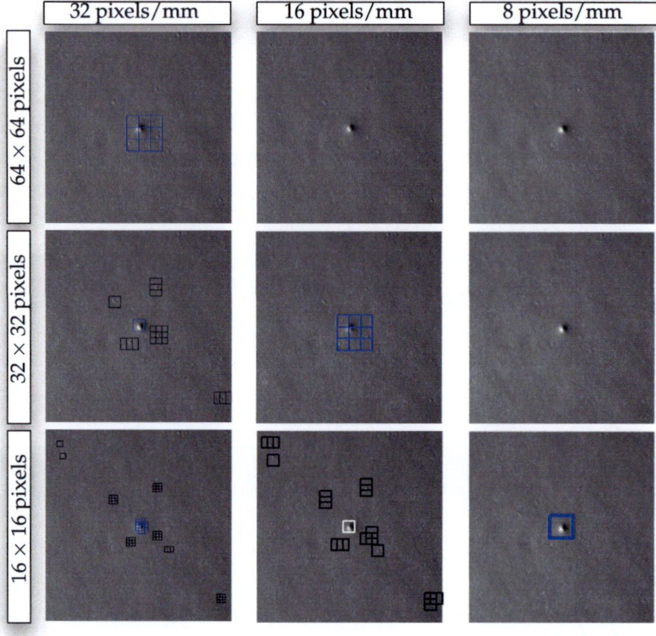

Figure 7.7: Classification results for a bubble and some peaks using different resolutions and window sizes. The color code of Fig. 7.1 was used.

This is fundamental for the viability of the proposed method in industrial applications. In this case, a small B makes not only the acquisition times shorter but also the acquisition hardware simpler. On the other hand, a small number of images in the series involves a loss of information. Additionally, the invariance against rotation decreases together with B. Using the same system of invariants as in the previous section, the classification rates and relative testing time are presented in Fig. 7.9 for different values of B and $\Delta\varphi = \Delta\omega$. As it can be seen, the best results are obtained for $B = 8$. Figure 7.10 shows the classification results for the examples of Fig. 7.5 and $B \in \{16, 8, 4\}$.

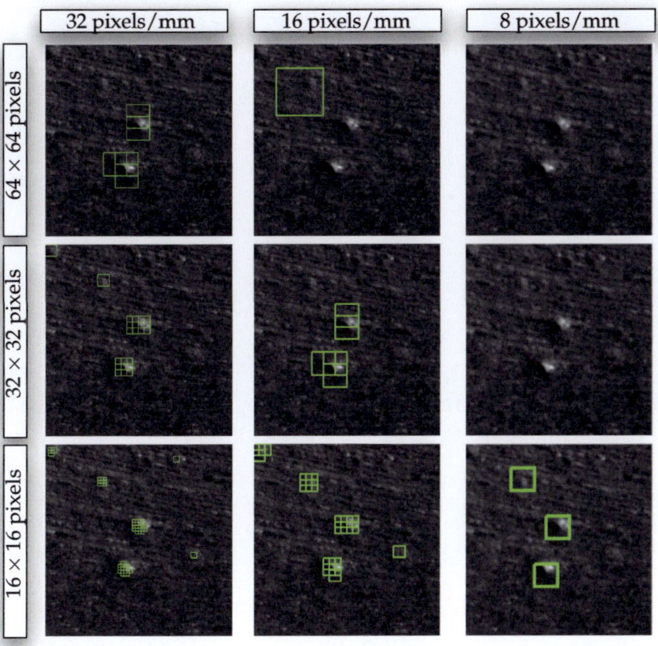

Figure 7.8: Classification results for 4 craters using different resolutions and window sizes. The color code of Fig. 7.1 was used.

7.3 Selection of z

In Eq. (5.40), the invariant feature $\tilde{\mathbf{f}}^l(\mathcal{S}_g)$ is modified by an exponential function z^d, where $z \in \mathbb{N}$ and d refers to the bins of the fuzzy histograms used to construct the feature. The function z^d improves the separability between features of different classes. The increment of the distance between features for values of $z > 1$ has been shown in Section 5.3.3 by calculating the Manhattan norm. However, this distance measure cannot be easily extended to the high dimensional spaces, in which the SVM performs the separaten of features according to given classes. In this case, the best way to show that z^d really improves the classifi-

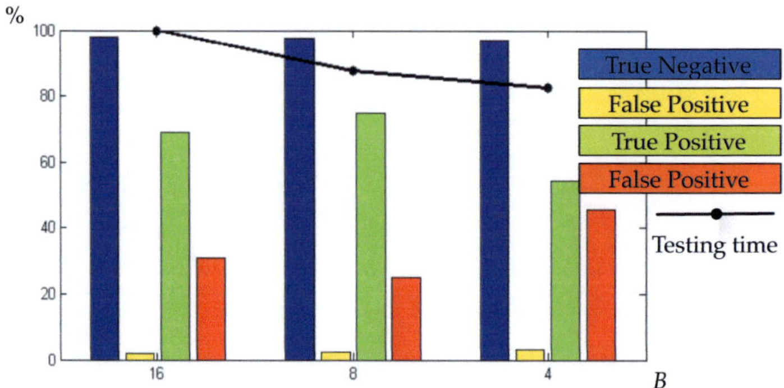

Figure 7.9: Classification rates tn, tp, fn and fp for $B \in \{16, 8, 4\}$. Testing time is normalized with regard to its respective maximum.

cation results is through an example. For this purpose, a small set of series of images showing fissures and bubbles, always on the same substrate texture, was chosen to train and test the SVM. For training and testing with the LOO method, only one invariant $\tilde{\mathbf{f}}^l(\mathcal{S}_g)$ ($L = 1$) with $\mathbf{w}^l = (0, 3, 0, 0, 0, 45°)$ is used, while the value of z is varied from 1 to 10. The classification results for 4 of the considered series of images are shown in Fig. 7.11. The first row shows the corresponding masks, whereas the second row presents the classification results delivered by the SVM for $z = 1$ (i.e., the trivial case). For $z = 1$, the selected invariant can only distinguish between no defect and fissure, but not between no defect and bubble. Using $z = 2$ (3rd row) a first bubble is detected (3rd column), but it is incorrectly classified as a fissure. That is, the function 2^d improves the distance between no defect and bubble, but not between bubble and fissure. Now, incrementing z to 4 (4th row) bubbles in the first and second column are correctly recognized. Further, for $z = 6$ and $z = 8$, results can be improved. Finally, all defects can be properly recognized with $z = 8$ except for one bubble that is classified as a fissure (6th row, 3rd column). For the given example, a greater z does not improve the results anymore.

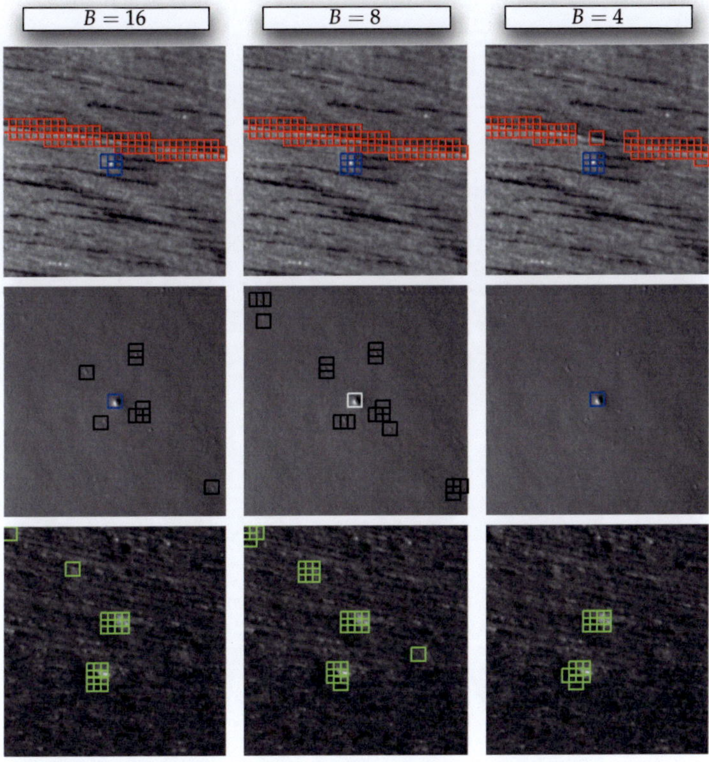

Figure 7.10: Classification results considering different values of B. The color code of Fig. 7.1 was used.

7.4 Evaluating classification results

In this section, the LOO method is applied to classify the complete set $\mathcal{L}$. Each series of images $\mathcal{S}_g \in \mathcal{L}$ is generated varying the illumination azimuth in steps of $\Delta\omega = 45°$, which results in series of eight images $B = 8$. As discussed above, for each image $g_{mnb} \in \mathcal{S}_g$ a resolution of 16 pixels/mm is used together with windows of 16×16 pixels. As kernel function for the SVM, a polynomial of degree 5 is used with $C = 1$ and a scale range of features given by $[-10, 10]$ (see Appendix B). Next, a system of invariants $\tilde{\mathcal{F}}_g$ must be selected, so that it is as close to completeness as possible (Eq. 5.38). For the given $\mathcal{L}$, $\tilde{\mathcal{F}}_g$ was selected experimentally by evaluating the resulting confusion matrices. The system composed by the invariants $\tilde{\mathbf{f}}^l(\mathcal{S}_g)$ listed in Table 7.1 with $L = 9$ is the

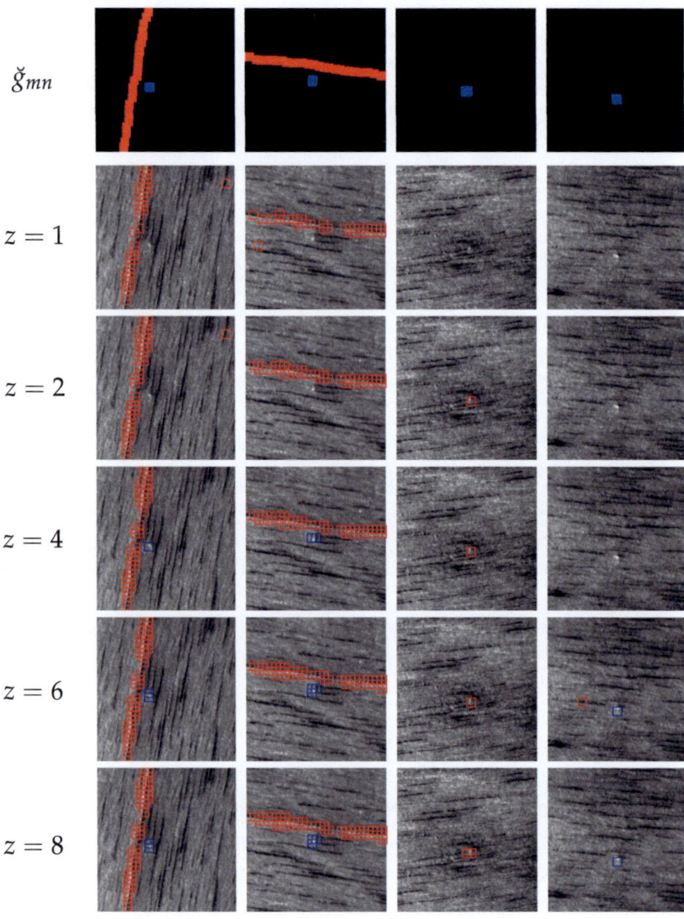

Figure 7.11: Classification results for a feature given by $\mathbf{w}^l = (0,3,0,0,0,45°)$ and $z \in \{1,2,4,6,8\}$. The color code of Fig. 7.1 was used.

Table 7.1: Parameters of feature set.

l	r_1	r_2	α	β	a	$\Delta\rho$	z	D
1	0	3	$0°$	$0°$	0	$45°$	10	5
2	0	10	$0°$	$0°$	0	$45°$	10	5
3	0	9	$0°$	$0°$	0	$45°$	10	5
4	0	1	$0°$	$0°$	0	$45°$	10	5
5	8	12	$0°$	$0°$	0	$45°$	10	5
6	6	6	$0°$	$180°$	0	$45°$	10	5
7	9	9	$0°$	$180°$	0	$45°$	10	5
8	1	1	$0°$	$180°$	0	$45°$	10	5
9	4	6	$0°$	$0°$	2	$45°$	10	5

one that yields the best results for $\mathcal{L}$. The confusion matrix for the complete set $\mathcal{L}$ is presented in Table 7.2. Appendix C shows the classification results, together with the first image of each series and the corresponding mask.

From the confusion matrix, the classification rates can be calculated: $tn = 99.68\%$, $fn = 44.10\%$, $fp = 0.05\%$, $tc = 50.71\%$ and $td = 5.19\%$ ($tp = 55.90\%$). For the given training conditions, very few false positives have been obtained (the rate of false alarms is very low), but there is a 44.10% false negatives. However, this does not mean that 44.10% defects are not detected at all, but just that as many windows $\mathcal{W}_g^w$ with $c > 0$ have been classified as no defect ($\hat{c} = 0$). To reflect the classification rates of defects instead of those of windows, the next section proposes a method to adapt the confusion matrix.

7.4.1 Classification rates for defects

The previous confusion matrix CM is rather a pessimistic approach to evaluate the classification performance. This is because the presented confusion matrix evaluates not only the recognition of defects, but also their segmentation in comparison with the mask. This is not meaningful because the defect detection and classification is more relevant than its exact segmentation. Moreover, during the mask generation, no special care is taken in perfectly delimiting defect contours. For these rea-

Table 7.2: Confusion matrix.

	no defect $c = 0$	crater $c = 1$	fissure $c = 2$	bubble $c = 3$	pin-holing $c = 4$	pickle $c = 5$	peak $c = 6$
no defect $\hat{c} = 0$	99.6814	38.4516	24.5540	39.3750	66.6667	33.1633	62.3656
crater $\hat{c} = 1$	0.0881	60.3871	0	0	0	2.0408	0
fissure $\hat{c} = 2$	0.0856	0.5161	73.8720	0	1.1111	1.0204	0
bubble $\hat{c} = 3$	0.0340	0.2581	0	53.7500	0	6.1224	1.0753
pin-holing $\hat{c} = 4$	0.0428	0	0.8395	0	28.8889	2.5510	4.3011
pickle $\hat{c} = 5$	0.0441	0.3871	0.3148	6.8750	3.3333	55.1020	0
peak $\hat{c} = 6$	0.0239	0	0.4197	0	0	0	32.2581

sons, if at least one window on a defect is correctly classified, the result can be considered to be satisfactory. In this case, the confusion matrix represents the classification of defects as a whole and not as a superposition of windows $\mathcal{W}_g^w$. Figure 7.12 schematizes through an example the proposed modification of the confusion matrix. In this example, a defect has been identified with 6 windows and labeled as class 1. Three possible classification results are presented in the figure: the first one coincides perfectly with the mask, the second classifies correctly only 2 windows on the defect and the last one recognizes 3 windows but classifies only two correctly. For practical applications, these three results indicate the presence of a defect of class 1 on the surface. Therefore, all of them can be considered to be correct. In order to reflect this fact in the confusion matrix, the three results are corrected to coincide with the mask. Then, a new confusion matrix CM' is calculated.

For the confusion matrix CM in Table 7.2 the proposed modification leads to the matrix CM' presented in Table 7.3.

Now, the new classification rates are: $tn = 99.61\%$, $fn = 16.82\%$, $fp = 0.06\%$, $tc = 78.32\%$ and $td = 4.86\%$ ($tp = 83.18\%$). These im-

Table 7.3: Confusion matrix CM'.

	no defect $c = 0$	crater $c = 1$	fissure $c = 2$	bubble $c = 3$	pin-holing $c = 4$	pickle $c = 5$	peak $c = 6$
no defect $\hat{c} = 0$	99.6122	11.2258	6.2959	5	30	0	48.3871
crater $\hat{c} = 1$	0.0693	88.7742	0	0.6250	0	4.0816	0
fissure $\hat{c} = 2$	0.1234	0	93.1794	0	0	0	0
bubble $\hat{c} = 3$	0.0693	0	0	88.1250	0	3.5714	0
pin-holing $\hat{c} = 4$	0.0478	0	0	0	65.5556	0	9.6774
pickle $\hat{c} = 5$	0.0302	0	0	5	4.4444	92.3469	0
peak $\hat{c} = 6$	0.0478	0	0.5247	1.25	0	0	41.9355

provements in fn and tp are due to the fact that a considerable percentage of false negatives in CM corresponds to discrepancies in segmenting defects with respect to the mask. As can be noticed in Appendix C, the segmentation differs mostly at defect borders. However, this does not affect the recognition task.

Let now CM' be analyzed class per class. The first column of CM' ($c = 0$) shows that the false alarm rate is fairly low. The defect that causes more false positives is the fissure ($c = 2$). This can be explained by its similarity to wood fiber. However, the false alarm rate is still not significant: 0.12%. The second column of CM' shows the classification of craters. This defect is correctly classified in 88.77% of the cases. The remaining craters are not detected at all. Comparing CM and CM', it can be deduced that crater edges are sometimes confused with fissures, bubbles and pickles but never with pin-holings and peaks. This can be associated with the similar profiles of these defects showed in Fig. 7.1. From the third column of CM', the fissure is the defect class that is most of the times correctly recognized. In the case of bubbles ($c = 3$), the correct recognition is around 88.13% of the cases. Further, bubbles are

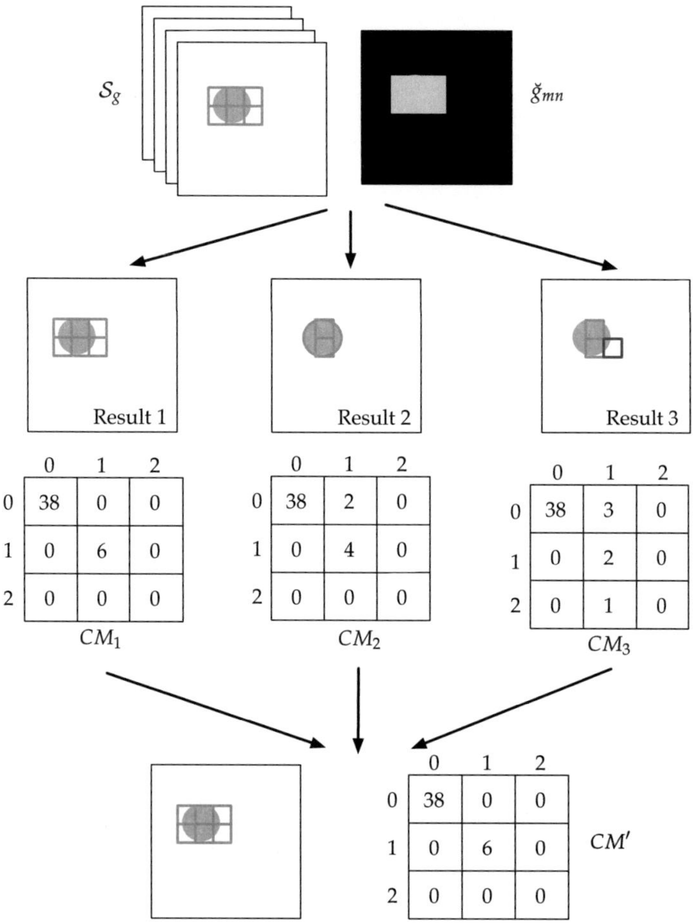

Figure 7.12: Modification of the confusion matrix to disregard segmentation discrepancies between the mask and the classification. For this example, the matrices CM are constructed with absolute values, where the total number of windows is given by $W = 44$.

confused with pickles in 5% of the cases. However, even industry experts have problems to distinguish between these two defect classes. A bubble and a pickle are characterized by elevations in the varnish film produced respectively by air and by foreign particles. Then, when a pickle is caused by a tiny particle, e.g., dust, it can look very similar to a bubble. In this case, the difference between these defects can only be appreciated under a microscope. Pin-holings ($c = 4$) are recognized only in 65.56% of the cases. However, this low classification rate can be compensated by incrementing the number of samples containing pin-holings in $\mathcal{L}$. The fifth column of CM' represents the classification performance with regard of pickles. This class of defects shows a high true classification rate of 92.35%. The last class considered, peaks, gives the highest false negative rate. Less than one half of peaks in $\mathcal{L}$ is correctly recognized. The reason of this poor classification rate can be attributed to the extremely small size of these defects. Nevertheless, this can be improved increasing the number peak samples in $\mathcal{L}$.

In conclusion, the proposed method gives a low classification performance only for the classes $c = 4$ and $c = 6$ (pin-holings and peaks). However, as mentioned before, this can be solved by introducing new samples of these defects into $\mathcal{L}$. For the rest of the classes, the classification performance is always around 90%. The images in Appendix C show that the different wood textures considered in $\mathcal{L}$ (Fig. 7.2) have no influence on the classification. Thus, the developed method is insensitive to the substrate texture. As expected, rotation and translation transformations (2D Euclidean motion) do not alter classification results.

8 Conclusions

In this thesis, a novel method was proposed to automatically detect and classify varnish defects on wood surfaces. These defects are responsible for a big economic wastage in industry. Additionally, the varnishing process involves the repeated use of dangerous chemicals and huge amounts of energy. Hence, the late detection of varnish defects yields extra pollution, because damaged pieces receive additional unneeded coating films. In order to avoid unnecessary pollution and economic loss, the proposed detection and classification method is thought to be performed between the different varnish applications. Different varnish defects can have different reasons, so that classifying defects allows identifying their causes.

The automated detection and classification of varnish defects on wood surfaces is a non-trivial problem and cannot be solved with standard techniques. For instance, varnish defects cannot be distinguished under diffuse illumination and are only partially visible under directional light. In addition, these defects are very small in comparison with the inspected surface. Moreover, if the varnish film is transparent, the wood texture can mask defects and/or cause false detections. Another issue that arises with regard to varnish defects is that they have a very similar appearance, so that it is very difficult to differentiate them from one another.

The proposed detection and classification method for varnish defects is based on series of images, which are obtained under variable illumination directions. A novel technique was presented to construct invariant features based upon these series of images. This involves the combination of integral invariants and fuzzy histograms together with a kernel function specifically designed for this problem. The resulting features are invariant against 2D rotation and translation (i.e., against 2D Euclidean motion). For a rotation, the surface relief interacts with the directional light producing local intensity changes in the images of the series. This was contemplated in the formulation of the 2D rotation transformation in order to extend it for series of images. As a result, the invariants consider not only geometrical transformations, but also the influence of directional light on the inspected surface. An extensive evaluation of the invariance properties of the proposed features was further presented in relation to the used illumination directions. As expected, series of images generated under a higher number of illumination directions (i.e., more images in the series) behave more robustly against 2D Euclidean motion. However, larger series of images increase the computational cost, so that a trade-off between robustness and running time must be found. Additionally, the robustness of the invariance properties was measured against different kernel function parameters, so as to determine their most suitable value ranges.

The obtained features are then classified by a Support Vector Machine (SVM). This classifier was selected because of its good generalization ability. The classification process is accomplished locally by the SVM, i.e., this is successively performed in smaller windows. The size of the windows is strongly related to the used image resolution and has a big influence on the classification results. The selection of the most appropriate combination of window size and image resolution has been thoroughly analyzed in this thesis with regard to both computational cost and classification rates.

For evaluating the classification results, the six most frequent defects classes were considered on different wood textures. The classification has been evaluated on a set of 85 series of images through the Leave One Out (LOO) method. Confusion matrices were constructed to perform a quantitative analysis of classification results per class. A rate of around 90% correct classifications is achieved for most of the considered defect classes. Further, this can be bettered by improving the training set. On the other hand, the false alarm rate is negligible for all classes, which demonstrates the tolerance of the proposed method regarding different wood texture.

In conclusion, the proposed method to detect and classify varnish defects on wood surfaces results in a two-dimensional invariant feature set that exhibits high discriminability and good generalization characteristics at the same time. Moreover, this method can be easily adapted to inspect other kind of surfaces that also require to be examined under variable illumination directions.

A Series of images

This appendix presents one series of images for each defect class.

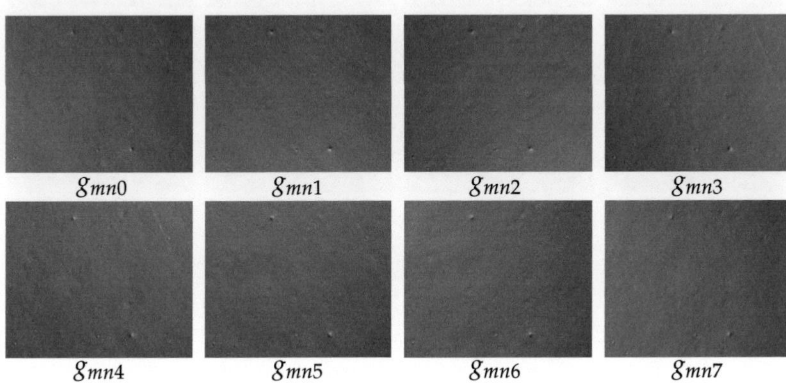

Figure A.1: Series of images of peaks and bubbles, $B = 8$.

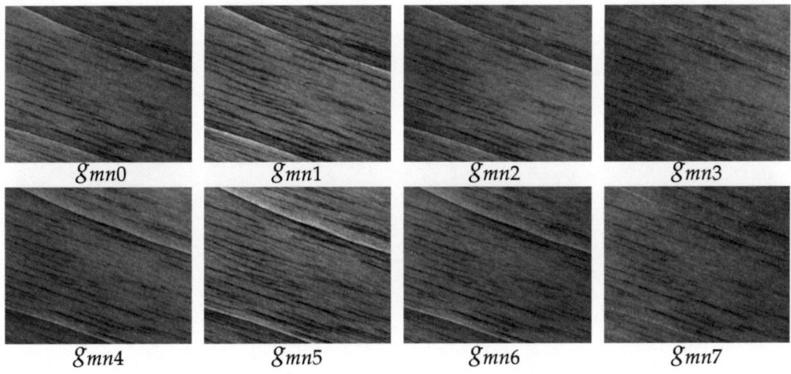

Figure A.2: Series of images of fissures, $B = 8$.

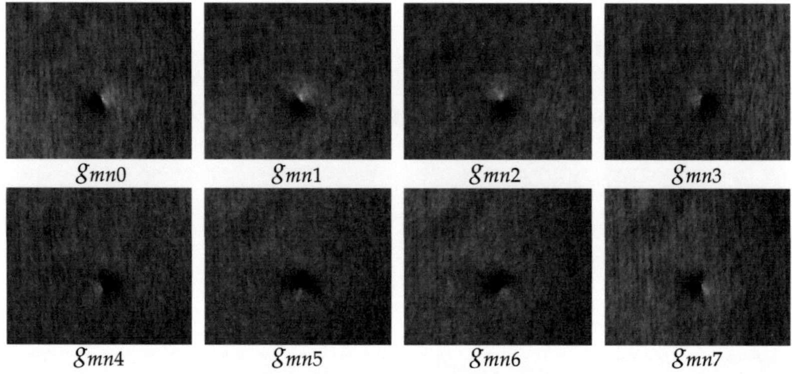

Figure A.3: Series of images of a crater, $B = 8$.

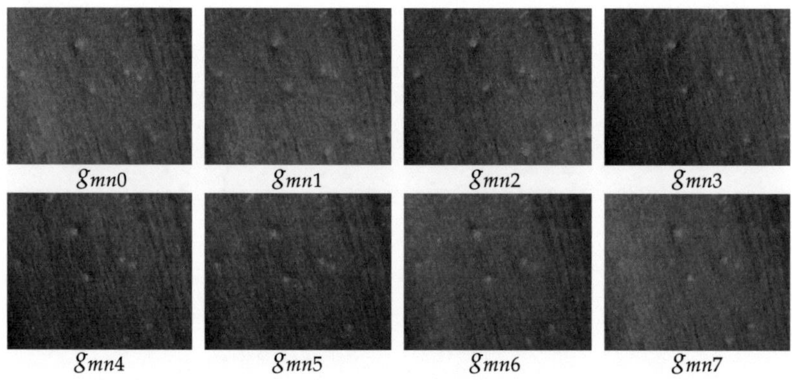

Figure A.4: Series of images of pin-holings, $B = 8$.

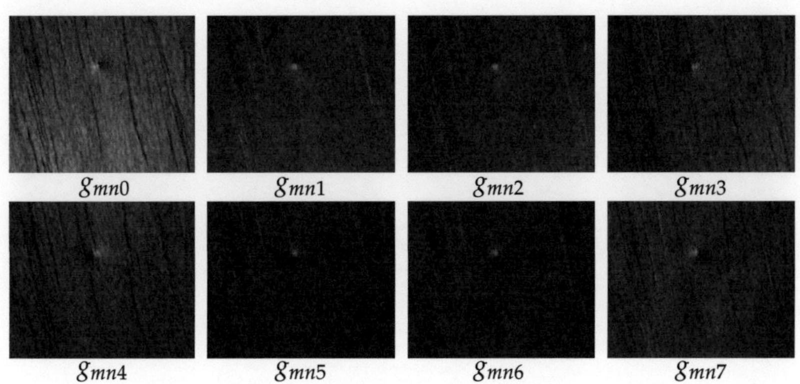

Figure A.5: Series of images of a bubble, $B = 8$.

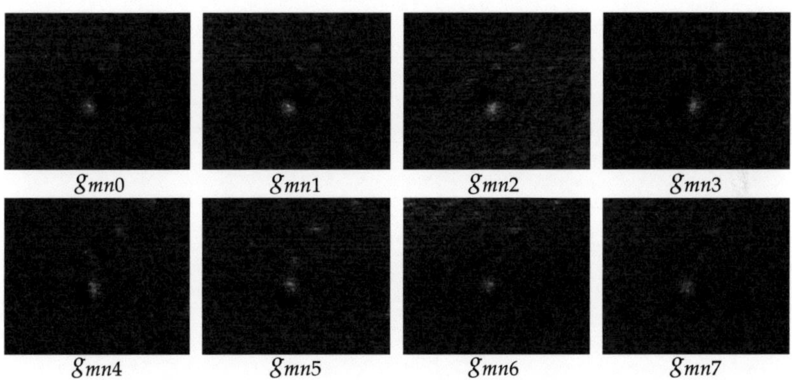

Figure A.6: Series of images of pickles and a crater, $B = 8$.

B Support Vector Machines

This Appendix presents an introduction to Support Vector Machines (SVM) based on [Bur98]. Next section introduces the concept of structural risk minimization, while the second and third sections describe the linear and non-linear SVM.

B.1 Structural risk minimization

The SVM were developed by Vapnik in 1995 and its formulation involves the Structural Risk Minimization (SRM) principle. The SRM is an inductive principle for selecting a model to learn from a set of finite training data. SRM minimizes an upper bound on the expected risk and has been shown to be superior to the traditional Empirical Risk Minimization (ERM). ERM is employed by conventional neuronal networks and minimizes the error on the training data. Furthermore, the use of SRM gives the SVM a greater ability to generalize [Gun98]. For a given learning task, with a finite amount of training data, the best generalization performance is achieved if the optimal balance is found between the quality of the particular training set and the ability to learn any training set without error. This latter ability is called *capacity* [Bur98].

The SVM classifies features considering only two classes iteratively, i.e., it analyzes the separation of one class against all the other classes together. Thus, although for this thesis $|\mathcal{C}| = 7$ holds, the presented discussion is performed considering $\mathcal{C} = \{-1, 1\}$, where $c = 1$ means one class (e.g., no defect) and $c = -1$ all the other classes (e.g., all kinds of defects, craters, bubbles, etc.). As defined in Section 6.1, the training data set is given by $\{(\tilde{\mathcal{F}}_{g^s w}, c_{g^s w})\}$, where $s \in \{1, \ldots, S\}$ and $w \in \{1, \ldots, W\}$. It is assumed that there exists some unknown probability distribution $P(\tilde{\mathcal{F}}, c)$ from which these training data are extracted. To simplify the notation, the training data set is denoted by $\{(\tilde{\mathcal{F}}_u, c_u)\}$, where $u \in \{1, \ldots, U = S \cdot W\}$.

Now, a *machine* is necessary whose task is to learn the mapping $\tilde{\mathcal{F}}_u \mapsto c_u$. This machine is defined by the set of all possible mappings

$\tilde{\mathcal{F}} \mapsto \check{c}(\tilde{\mathcal{F}}, \zeta) = c$, where ζ is an adjustable parameter. The machine is assumed to be deterministic: for a given input $\tilde{\mathcal{F}}_u$ and choice of ζ, it always gives the same output $\check{c}(\tilde{\mathcal{F}}_u, \zeta) = c_u$. A particular choice of ζ generates a *trained machine*. The expected test error for a trained machine is therefore:

$$R(\zeta) = \int \frac{1}{2} |c - \check{c}(\tilde{\mathcal{F}}, \zeta)| \, \mathrm{d}P(\tilde{\mathcal{F}}, c). \tag{B.1}$$

The quantity $R(\zeta)$ is called *expected risk* and presents the problem that $P(\tilde{\mathcal{F}}, c)$ is unknown. However, for a fixed finite amount of training data given by U, it is possible to find an approximation calculating the *empirical risk*:

$$R_{\mathrm{emp}}(\zeta) = \frac{1}{2U} \sum_{u=1}^{U} |c_u - \check{c}(\tilde{\mathcal{F}}_u, \zeta)|. \tag{B.2}$$

In Eq. (B.2), no probability distribution appears. $R_{\mathrm{emp}}(\zeta)$ is a fixed number for a particular choice of ζ and for a particular training set $\{(\tilde{\mathcal{F}}_u, c_u)\}$. The quantity $\frac{1}{2} |c_u - \check{c}(\tilde{\mathcal{F}}_u, \zeta)|$ is called *loss*. For the considered $\mathcal{C} = \{-1, 1\}$, the loss can only take the values 0 and 1. For a probability given by $1 - \eta$, the following bound holds:

$$R(\zeta) \leq R_{\mathrm{emp}}(\zeta) + \sqrt{\frac{\lambda(\log(2U/\lambda) + 1) - \log(\eta/4)}{U}}, \tag{B.3}$$

where λ is a non-negative integer called *Vapnik Chervonekis dimension* (VC). The VC value λ is a measure of the machine capacity. The right hand side of Eq. (B.3) is called *risk bound*. The second term of the risk bound is called *VC confidence*. This bound assumes that the training and test data are drawn independently according to some $P(\tilde{\mathcal{F}}, c)$. If λ is known, the risk bound can be computed. Thus, given several different learning machines (i.e., different families of functions $\check{c}(\tilde{\mathcal{F}}, \zeta)$) and choosing a fixed, sufficiently small η, the lowest upper bound of the expected risk can be found by selecting the machine that minimizes the right hand side of Eq. (B.3).

The VC dimension is a property of the set of functions $\{\check{c}(\zeta)\}$. Considering as before $\check{c}(\tilde{\mathcal{F}}, \zeta) \in \{-1, 1\}$, a training data set of U elements can be classified in 2^U possible ways. If for each of the 2^U possible classification constellations, there exists a member of $\{\check{c}(\zeta)\}$ that correctly assigns the classes, it is said that this data set is *shattered* by $\{\breve{c}(\zeta)\}$. The VC dimension for a set of functions $\{\breve{c}(\zeta)\}$ is defined as the maximal number of training data elements that can be shattered by $\{\breve{c}(\zeta)\}$. If the VC dimension is λ, then there exists at least one set of λ elements that can be shattered by $\{\breve{c}(\zeta)\}$. However, in general, $\{\breve{c}(\zeta)\}$ cannot shatter every set of λ elements. Figure B.1 shows an example of three elements in $\mathbb{R}^2$ (8 possible labelings) and a function family $\{\breve{c}(\zeta)\}$ constituted by oriented lines ($\zeta = (\mathbf{w}, b)$, where $\mathbf{w}$ is the line slope and b^1 the line bias). The family set of oriented lines has a VC dimension $\lambda = 3$ (4 elements in $\mathbb{R}^2$ cannot be shattered by oriented lines). The VC confidence term

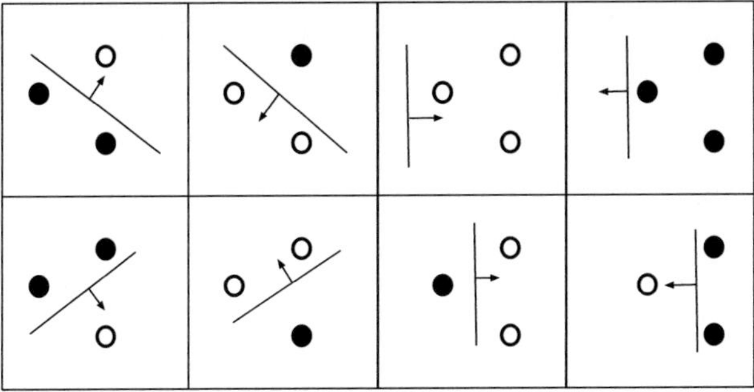

Figure B.1: Set of 3 training data elements. The white and black circles represent features of class $c = 1$ and $c = -1$ respectively. The machine is given by the set of all oriented lines $\{\breve{c}(\zeta = (\mathbf{w}, b)\}$. The resulting 2^3 possible classification constellations can be shattered by this set of functions.

depends on the chosen class of functions (e.g., oriented lines, parabolas, etc.). On the other hand, the empirical and expected risk depend on the one particular function chosen to train (i.e., the selected value of ζ). The SRM describes a general model for capacity control. By finding the function subset that minimizes the bound on the expected risk, the SRM achieves a trade-off between hypothesis space complexity (VC dimension) and the quality of fitting training data (empirical error). This can be done by dividing the entire set of functions $\{\breve{c}(\zeta)\}$ into nested subsets

---

[1]In this appendix, b denotes a line bias (it has not relation with the illumination directions used to generated the series of images).

according to their VC dimensions and training to determine which machines from each subset minimize the empirical risk. Then, the trained machine is selected for which the summation of the empirical risk and VC confidence is minimal (see Fig. B.2).

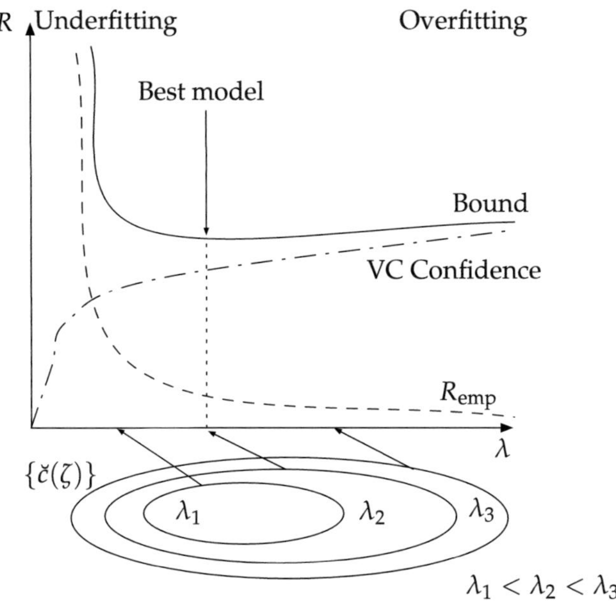

Figure B.2: Structural risk minimization.

B.2 Linear Support Vector Machine

B.2.1 The separable case

Let the machine be linear trained on separable data $\zeta = (\mathbf{w}, b)$. Let $\mathcal{Y}$ be some hyperplane, which separates the positive from the negative data (see Fig. B.3). The features $\tilde{\mathcal{F}}_u$ which lie on $\mathcal{Y}$ satisfy:

$$\langle \mathbf{w}, \tilde{\mathcal{F}}_u \rangle + b = 0,$$

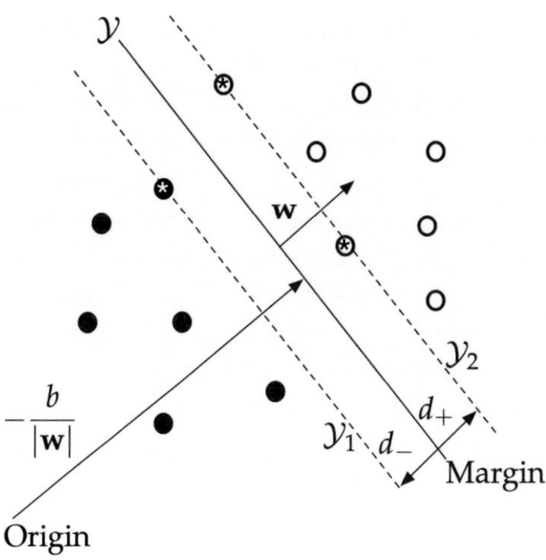

Figure B.3: Linear-separable data.

where $\mathbf{w}$ is the normal to $\mathcal{Y}$, $|b|/\|\mathbf{w}\|$ is the perpendicular distance from $\mathcal{Y}$ to the origin and $\langle \cdot, \cdot \rangle$ represents the inner product. A data set is said to be optimally separated by the hyperplane $\mathcal{Y}$, if it is separated without error and the distance between the closest data to $\mathcal{Y}$ is maximal. Let d_+ and d_- be respectively the shortest distance from $\mathcal{Y}$ to the closest positive and negative data. The *margin* of the separating hyperplane is defined as $d_- + d_+$ (see Fig. B.3). For the linearly separable case, the SVM algorithm simply looks for the separating plane $\mathcal{Y}$ with largest margin. Without loss of generality, the canonical hyperplane can be considered, where the parameters $\mathbf{w}$ and b are constrained by:

$$\langle \mathbf{w}, \tilde{\mathcal{F}}_u \rangle + b \;\leq\; -1 \quad \text{for } c_u = -1, \tag{B.4}$$

$$\langle \mathbf{w}, \tilde{\mathcal{F}}_u \rangle + b \;\geq\; +1 \quad \text{for } c_u = +1. \tag{B.5}$$

This can be combined into one inequality:

$$c_u \cdot \left(\langle \mathbf{w}, \tilde{\mathcal{F}}_u \rangle + b \right) - 1 \geq 0 \quad \forall u. \tag{B.6}$$

The data, for which the equality in Eq. (B.4) holds, lie on the hyperplanes $\mathcal{Y}_1$ with normal $\mathbf{w}$ and perpendicular distance from the origin

$|1 - b|/\|\mathbf{w}\|$. Similarly, the data, for which the equality in Eq. (B.5) holds, lie on the hyperplane $\mathcal{Y}_2$, whose perpendicular distance from the origin is $|-1 - b|/\|\mathbf{w}\|$. Hence, $d_- = d_+ = 1/\|\mathbf{w}\|$ and therefore, the margin is $2/\|\mathbf{w}\|$. Then, the hyperplane that optimally separates the features is the one that minimizes $\frac{1}{2}\|\mathbf{w}\|^2$ under the constraint given in Eq. (B.6). Intuitively, this hyperplane depends only on the training data for which the equality in Eq. (B.6) holds. These data are called support vectors (support vectors are denoted in Fig. B.3 with $*$). The minimization of $\frac{1}{2}\|\mathbf{w}\|^2$ is equivalent to implement the SRM principle, that is, to minimize the upper bound on the VC dimension. A Lagrangian formulation can be used to minimize $\frac{1}{2}\|\mathbf{w}\|^2$ subject to Eq. (B.6). Introducing positive Lagrange multipliers ψ_u the following expression is obtained:

$$\ell(\mathbf{w}, b, \psi) = \frac{1}{2}\|\mathbf{w}\|^2 - \sum_{u=1}^{U} \psi_u c_u \left(\langle \mathbf{w}, \tilde{\mathcal{F}}_u \rangle + b\right) + \sum_{u=1}^{U} \psi_u. \tag{B.7}$$

Now, ℓ must be minimized with respect to $\mathbf{w}$ and b, while $\partial\ell/\partial\psi_u = 0$ for all $\psi_u \geq 0$. This is a convex quadratic programming problem, since the objective function is itself convex, and those data that satisfy the constraints also form a convex set. This means, that the dual problem can be equivalently solved (for a geometrical interpretation of the duality in SVM, see [BB00]). The dual formulation allows defining a function ℓ_D, which depends only on the Lagrange multipliers ψ_u and the data inner product. The dual problem is given by:

$$\max_{\psi}(\ell_D(\psi)) = \max_{\psi}(\min_{\mathbf{w}, b} \ell(\mathbf{w}, b, \psi)). \tag{B.8}$$

For each minimum of $\ell(\mathbf{w}, b, \psi)$ with respect to $\mathbf{w}$ and b, the corresponding gradients must vanish:

$$\frac{\partial\ell}{\partial\mathbf{w}} = 0 \quad \Rightarrow \quad \mathbf{w} = \sum_{u} \psi_u c_u \tilde{\mathcal{F}}_u, \tag{B.9}$$

$$\frac{\partial\ell}{\partial b} = 0 \quad \Rightarrow \quad \sum_{u} \psi_u c_u = 0. \tag{B.10}$$

Considering that $\|\mathbf{w}\|^2 = \mathbf{w}^{\mathrm{T}}\mathbf{w}$ and denoting $\mathbf{w}^{\mathrm{T}} = \sum_{u'} \psi_{u'} c_{u'} \tilde{\mathcal{F}}_{u'}$, replacing Eq. (B.9) and Eq. (B.10) into Eq. (B.7) gives:

$$\ell_D(\psi) = -\frac{1}{2} \sum_{u,u'} \psi_u \psi_{u'} c_u c_{u'} \langle \tilde{\mathcal{F}}_u, \tilde{\mathcal{F}}_{u'} \rangle + \sum_{u} \psi_u. \tag{B.11}$$

Now, ℓ_D must be maximized with respect to ψ_u subject to $\psi_u \geq 0$ and $\sum_u \psi_u c_u = 0$. The values ψ_u^* that maximize ℓ_D can be found by numerical methods. The solution gives $\psi_u^* \neq 0$ only for those $\tilde{\mathcal{F}}_u$ which are support vectors. Then, if the set of support vectors is denoted by $\mathcal{SV}$ the solution $\mathbf{w}^*$ is given by:

$$\mathbf{w}^* = \sum_{\mathcal{SV}} \psi_u^* c_u \tilde{\mathcal{F}}_u. \tag{B.12}$$

The solution b^* can be obtained using the KKT (Karush Kuhn Tucker) complementary condition $\psi_u \left(c_u \left(\langle \mathbf{w}, \tilde{\mathcal{F}}_u \rangle + b \right) - 1 \right) = 0$. As $\psi_u \neq 0$ only for $\tilde{\mathcal{F}}_u \in \mathcal{SV}$, b^* can be found solving this equation for some support vector.

With $\mathbf{w}^*$ and b^* the optimal separating hyperplane is defined, i.e., the SVM is trained. Then, for a given test feature $\tilde{\mathcal{F}}'_u$, it must be decided on which side of the boundary $\mathcal{Y}$ it lies. A hard classifier is defined by:

$$\hat{c}_u = \mathrm{sgn} \left(\langle \mathbf{w}^*, \tilde{\mathcal{F}}'_u \rangle + b^* \right). \tag{B.13}$$

B.2.2 Non-separable case

In this subsection, the previous results are extended to consider non-separable data. This can be performed using a function $\check{c}(\zeta)$, which is more complex than a hyperplane. Another possibility is to generalize the optimal separating hyperplane by relaxing, when necessary, the constraints given in Eq. (B.4) and Eq. (B.5) (Cortes and Vapnik 1995). With this purpose, a positive slack variable $\varsigma_u \geq 0$ is introduced in the constraints (see Fig. B.4):

$$\langle \mathbf{w}, \tilde{\mathcal{F}}_u \rangle + b \leq -1 + \varsigma_u \quad \text{for } c_u = -1, \tag{B.14}$$

$$\langle \mathbf{w}, \tilde{\mathcal{F}}_u \rangle + b \geq +1 - \varsigma_u \quad \text{for } c_u = +1. \tag{B.15}$$

The optimization problem consists now in minimizing the classification error as well as minimizing the bound on the classifier VC dimension. The natural way to assign an extra cost for errors is to change the objective function $\|\mathbf{w}\|^2/2$ to $\|\mathbf{w}\|^2/2 + C \sum_u \varsigma_u$, where C is a parameter

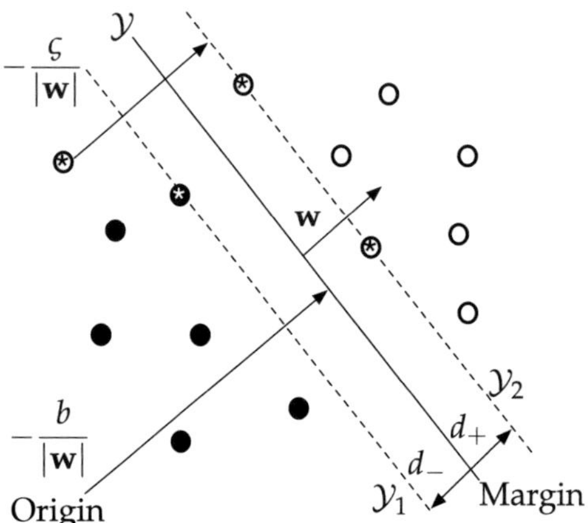

Figure B.4: Non-separable data.

to be chosen by the user. A larger C corresponds to assigning a higher penalty to errors. The primal Lagrangian can now be formulated as:

$$\ell(\mathbf{w},b,\psi,\varsigma,\gamma) \;=\; \frac{1}{2}\|\mathbf{w}\|^2 + C\sum_u \varsigma_u - \tag{B.16}$$

$$-\; \sum_u \psi_u \left(c_u \cdot \left(\langle\!\langle \mathbf{w},\tilde{\mathcal{F}}_u\rangle + b\right) - 1 + \varsigma_u\right) - \sum_u \gamma_u \varsigma_u,$$

where γ_u are the Lagrange multipliers introduced to force the positivity of ς_u. As before, $\ell(\mathbf{w},b,\psi,\varsigma,\gamma)$ has to be minimized with respect to $\mathbf{w}$, b and ς and maximized with respect to ψ_u and γ_u. Again, the dual problem is formulated:

$$\max_{\psi,\gamma}(\ell_D(\psi,\gamma)) = \max_{\psi,\gamma}(\min_{\mathbf{w},b,\varsigma} \ell(\mathbf{w},b,\psi,\varsigma,\gamma)). \tag{B.17}$$

For each minimum of $\ell(\mathbf{w},b,\psi,\varsigma,\gamma)$ with respect to $\mathbf{w}$, b and ς, the respective gradients must vanish:

$$\frac{\partial \ell}{\partial \mathbf{w}} = 0 \quad \Rightarrow \quad \mathbf{w} = \sum_u \psi_u c_u \tilde{\mathcal{F}}_u, \tag{B.18}$$

$$\frac{\partial \ell}{\partial b} = 0 \quad \Rightarrow \quad \sum_u \psi_u c_u = 0, \tag{B.19}$$

$$\frac{\partial \ell}{\partial \varsigma} = 0 \quad \Rightarrow \quad \psi_u + \gamma_u = C. \tag{B.20}$$

Introducing this results in ℓ_D:

$$\ell_D(\psi, \gamma) = -\frac{1}{2} \sum_{u,u'} \psi_u \psi_{u'} c_u c_{u'} \langle \tilde{\mathcal{F}}_u, \tilde{\mathcal{F}}_{u'} \rangle + \sum_u \psi_u. \tag{B.21}$$

Now, the solution is given by ψ^* and γ^*, which maximize ℓ_D, subject to the constrains $0 \leq \psi_u \leq C$ and $\sum_s \psi_u c_u = 0$. Then, $\mathbf{w}^*$ coincides with Eq. (B.12) and b^* can be obtained using the KKT condition:

$$\psi_u \left(c_u \left(\langle \mathbf{w}, \tilde{\mathcal{F}}_u \rangle + b \right) - 1 + \varsigma_u \right) = 0, \tag{B.22}$$

$$\gamma_u \varsigma_u = 0. \tag{B.23}$$

The combination of Eq. (B.23) and Eq. (B.20) shows that $\varsigma_u = 0$ if $\psi_u < C$. Then, using any training data for which $0 < \psi_u < C$, Eq. (B.22) can be used to compute b^*. The uncertain part of this approach is that the coefficient C has to be determined. This parameter introduces additional capacity control within the classifier and must be chosen to reflect the knowledge of noise on the training data.

B.3 Non-linear SVM

The methods presented above are now extended to the case where the decision function $\check{c}(\zeta)$ is a not a linear function. Boser, Guyon and Vapnik showed in 1992 that this generalization can be achieved in a straightforward way using an old trick (Aizerman 1964) and the fact that the training data appear in form of inner product in Eq. (B.21). Let $E : \mathbb{R}^L \mapsto \mathcal{N}$ be a mapping of the training data $\tilde{\mathcal{F}}_u$ to some other Euclidean space $\mathcal{N}$. Now, the training algorithm would depend on the data inner products in $\mathcal{N}$, that is, on $\langle E(\tilde{\mathcal{F}}_u), E(\tilde{\mathcal{F}}_{u'}) \rangle$. If there were a kernel function $\check{e}$ (which should not be confused with the kernel function used in the Haar integral) such that $\check{e}(\tilde{\mathcal{F}}_u, \tilde{\mathcal{F}}_{u'}) = \langle E(\tilde{\mathcal{F}}_u), E(\tilde{\mathcal{F}}_{u'}) \rangle$, then it would be only necessary to use $\check{e}$ in the training algorithm and E would not be necessary to be known explicitly. Then, replacing

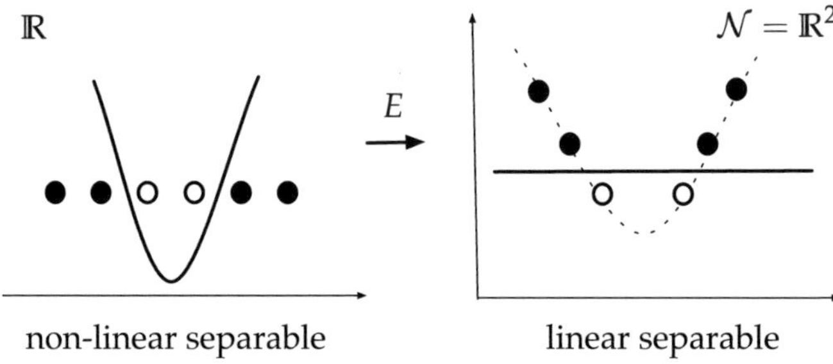

Figure B.5: A set of non-linear separable data in $\mathbb{R}$ can be linearly separated, if it is mapped into $\mathbb{R}^2$.

$\langle \tilde{\mathcal{F}}_u, \tilde{\mathcal{F}}_{u'} \rangle$ by $\breve{e}(\tilde{\mathcal{F}}_u, \tilde{\mathcal{F}}_{u'})$ everywhere in the training algorithm, an SVM for an $\mathcal{N}$ dimensional space is obtained. All the considerations of the previous sections hold, since still a linear separation is performed, but in a different space (see Fig. B.5). Now, a hard classifier can be defined as:

$$\hat{c}_u = \mathrm{sgn} \sum_{\mathcal{SV}} \psi_u c_u \, \breve{e}(\tilde{\mathcal{F}}_u, \tilde{\mathcal{F}}'_u) + b^*. \tag{B.24}$$

The pair $\{\mathcal{N}, E\}$ exists if the function $\breve{e}$ satisfies the Mercer's condition (Mercer 1909). A list of some valid kernel functions that satisfy Mercer's conditions are enumerated next:

- Polynomial function:

$$\breve{e}(\tilde{\mathcal{F}}_u, \tilde{\mathcal{F}}_{u'}) = \left(\langle \tilde{\mathcal{F}}_u, \tilde{\mathcal{F}}_{u'} \rangle + 1 \right)^{gr}.$$

- Gaussian Radial Basis function:

$$\breve{e}(\tilde{\mathcal{F}}_u, \tilde{\mathcal{F}}_{u'}) = \exp\left(-\frac{\|\tilde{\mathcal{F}}_u - \tilde{\mathcal{F}}_{u'}\|^2}{2\sigma^2} \right).$$

- Exponential Radial Basis function:

$$\breve{e}(\tilde{\mathcal{F}}_u, \tilde{\mathcal{F}}_{u'}) = \exp\left(-\frac{\|\tilde{\mathcal{F}}_u - \tilde{\mathcal{F}}_{u'}\|}{2\sigma^2} \right).$$

C Results

This appendix presents the classification results for the series of images $S_g \in \mathcal{L}$. The following pages show the first image g_{mn0} of the series together with the corresponding mask $\breve{g}_{mn}$ and the result of the classification $c_f(\tilde{\mathcal{F}}_g)$. The color code used for masks and classification results is introduced in Fig. 7.1. Comparing masks and classification results in this appendix gives a qualitative idea of the classification performance, while Chapter 7 presents the quantitive analysis of these results.

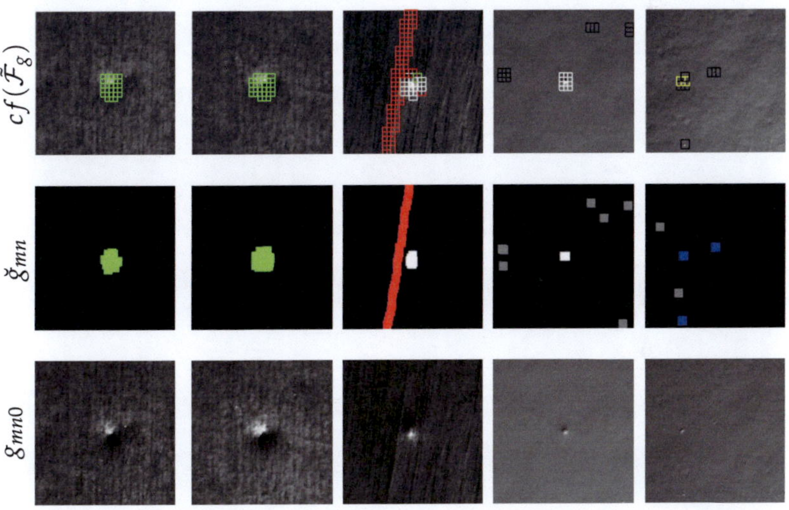

Results

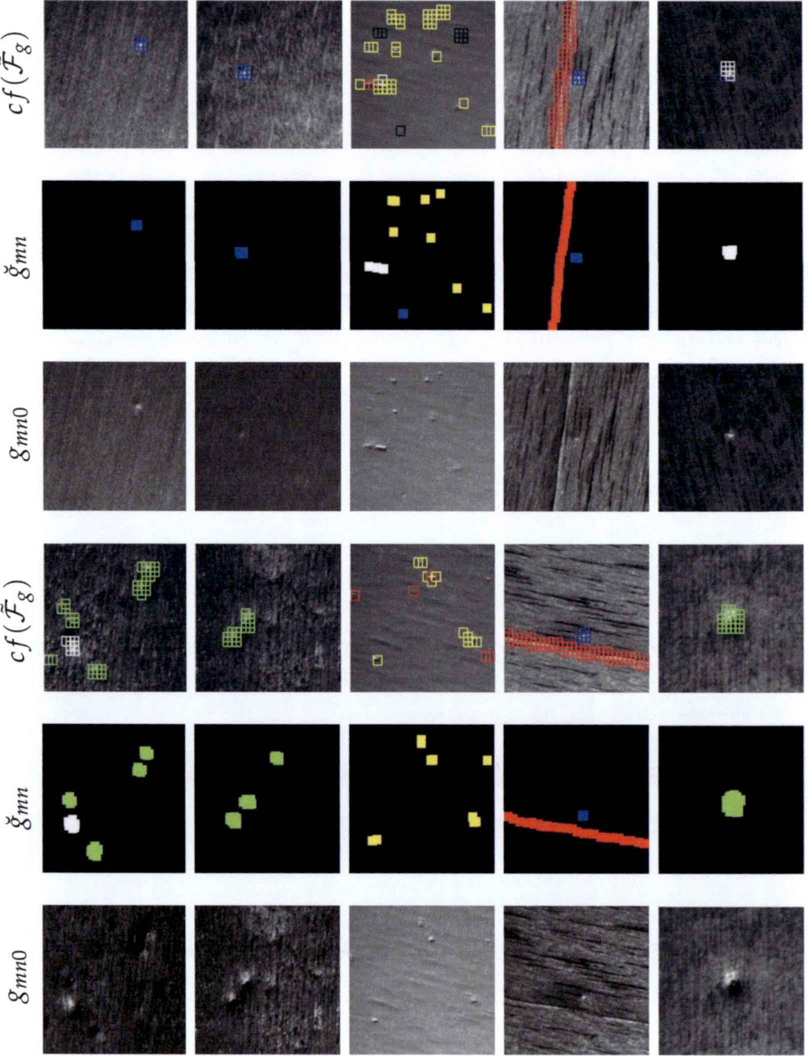

Results

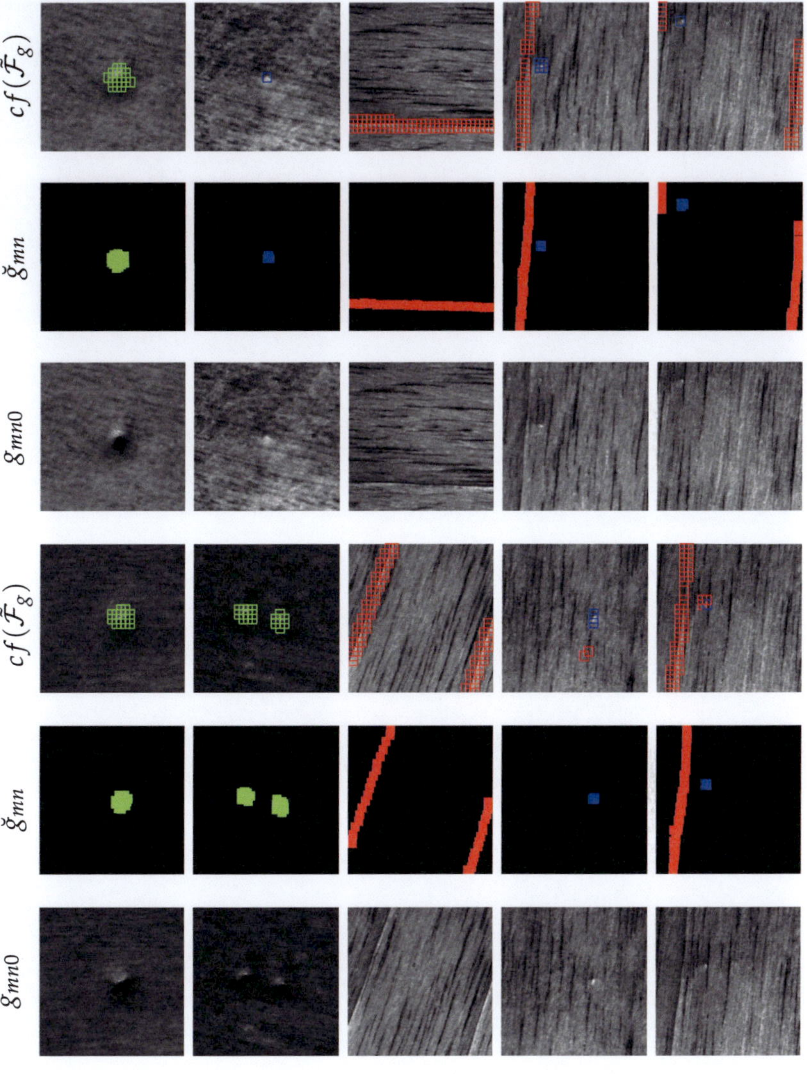

Results

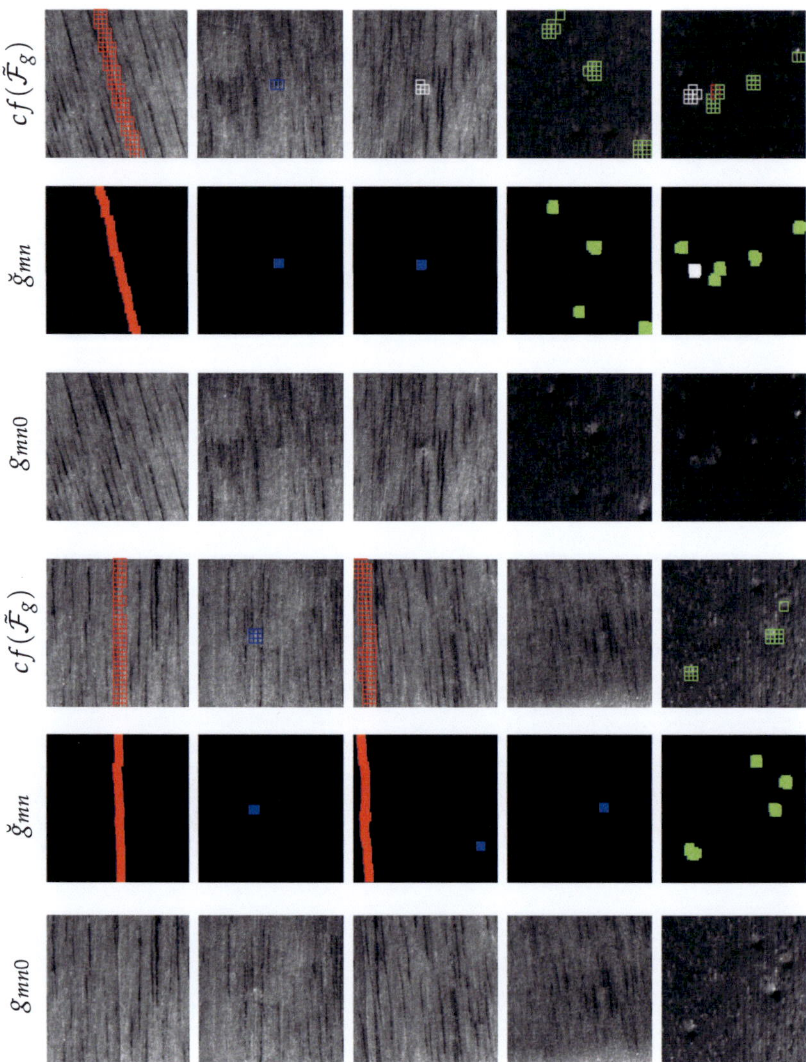

Results

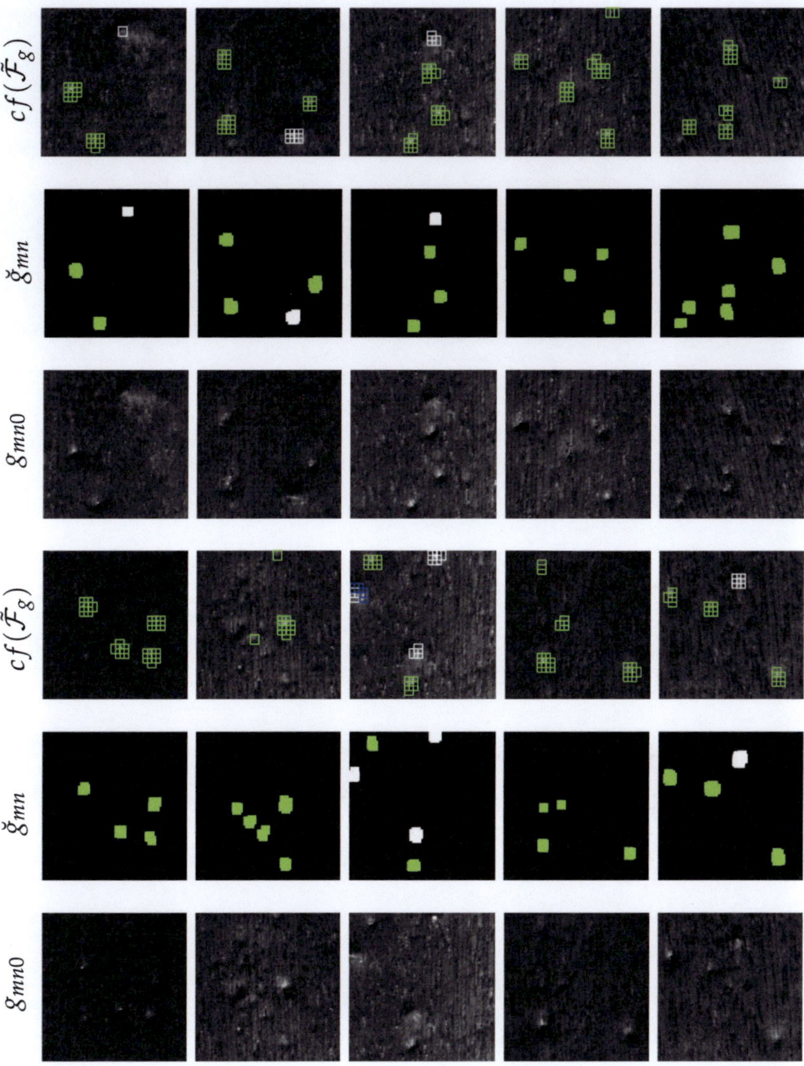

Results

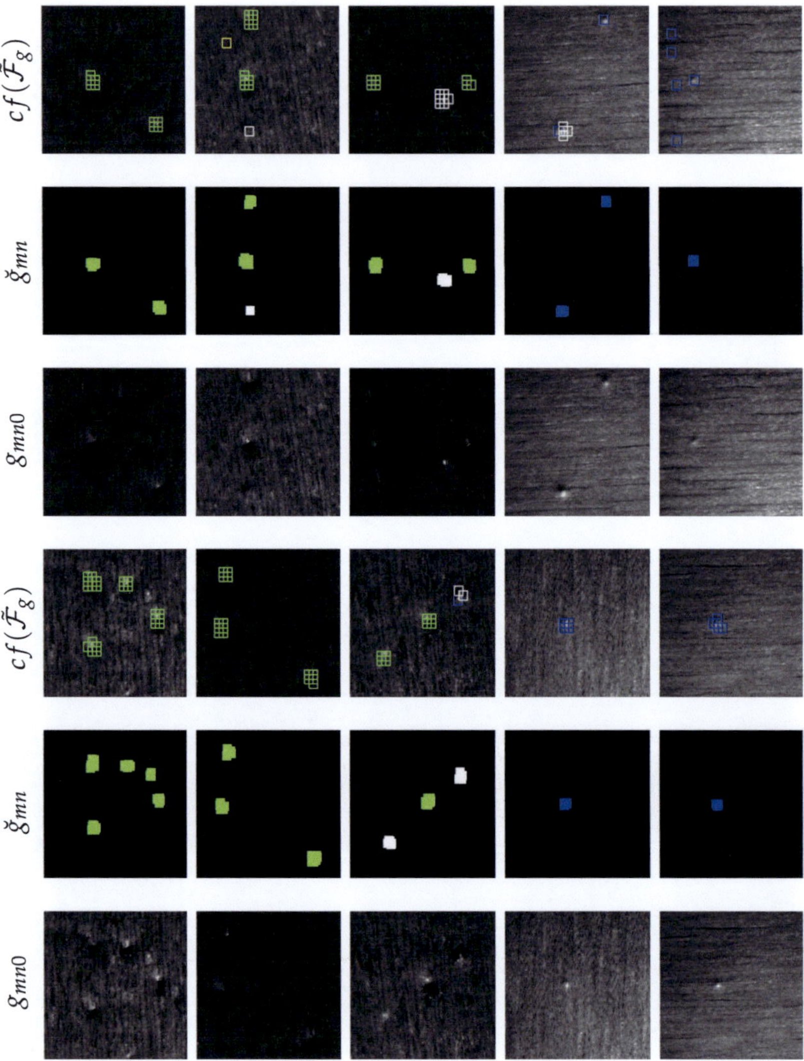

Results

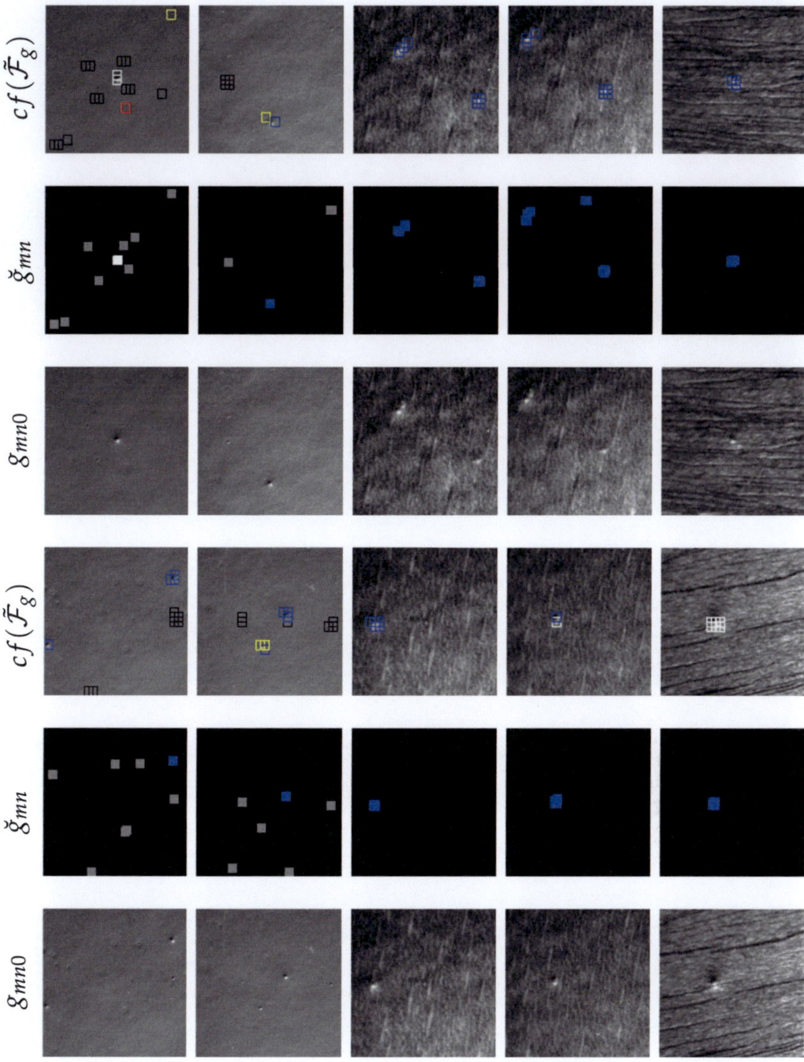

Results

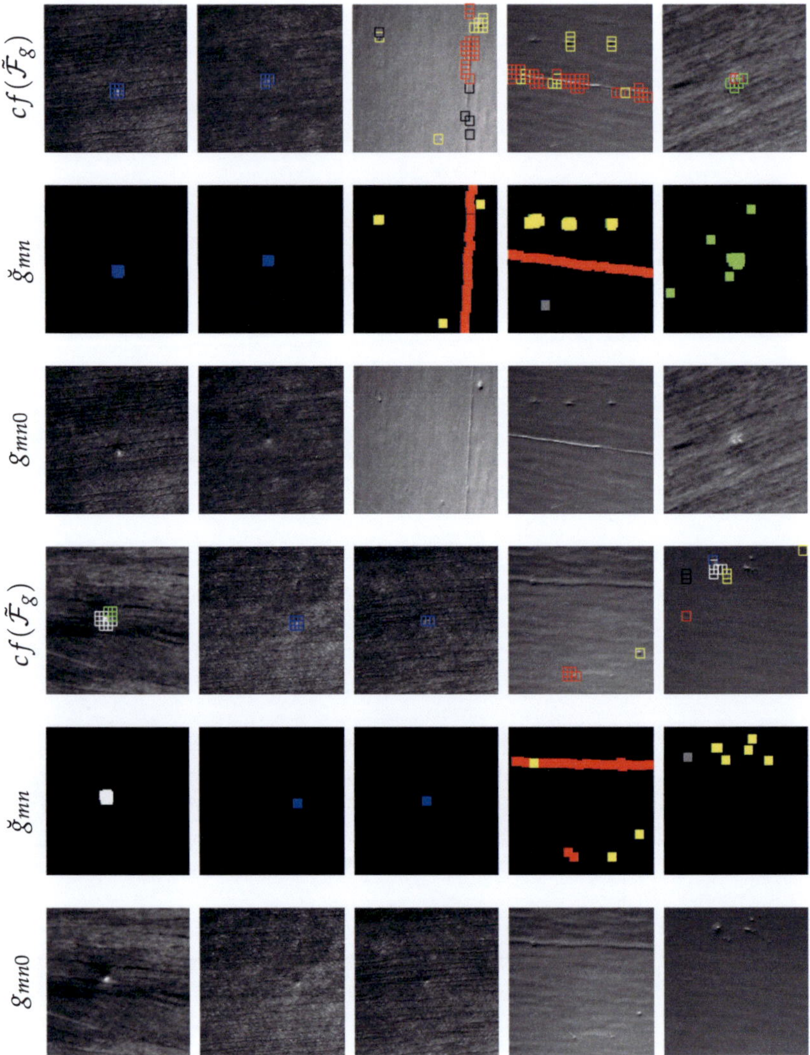

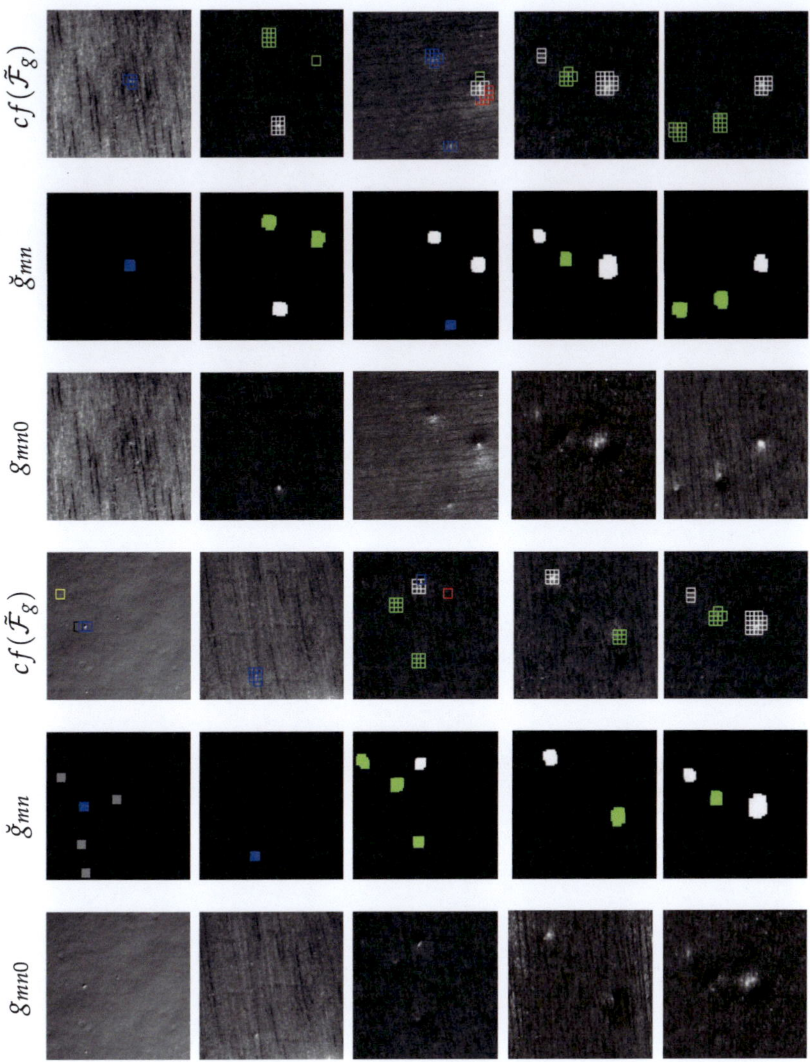

D Nomenclature

Table D.1: Principal symbols.

Symbol	Description
	Chapter 3: Introduction to pattern recognition
$\mathcal{O}$	Set of objects o in the real world
o	Object in the real world $o \in \mathcal{O}$
$\mathcal{C}$	Set of classes c
c	Class $c \in \mathcal{C}$
C_o	Object classification operator $C_o : \mathcal{O} \longrightarrow \mathcal{C}$
S	Measurement operator $S : \mathcal{O} \longrightarrow \mathcal{G}$
$\mathcal{G}$	Pattern space
$g(x)$	Pattern $g(x) \in \mathcal{G}$; $g := g(x)$
$\mathcal{X}$	Set of pattern variables
x	Pattern variable $x \in \mathcal{X}$
c_g	Pattern classification function $c_g : \mathcal{G} \longrightarrow \mathcal{C}$
$\mathcal{E}_c$	Equivalence class $\mathcal{E}_c \subset \mathcal{O}$
$f(g)$	Feature $f : \mathcal{G} \longrightarrow \mathcal{F}$; $g \mapsto f(g)$
$\mathcal{F}$	Feature space $f(g) \in \mathcal{F}$
c_f	Feature classification function $c_f : \mathcal{F} \longrightarrow \mathcal{C}$
$\mathcal{E}_g$	Equivalence class $\mathcal{E}_g \subset \mathcal{G}$
	Section 3.1: Classification by extracting invariants
$T(t)$	Induced transformation on $\mathcal{G}$; $T : \mathcal{G} \times \mathcal{T} \longrightarrow \mathcal{G}$
$\mathcal{T}$	Set of transformations parameters t
t	Transformation parameters $t \in \mathcal{T}$
$T(\mathcal{T})$	Transformation set $T(t) \in T(\mathcal{T})$
p	Pattern taken as prototype $p \in \mathcal{G}$
$\mathcal{E}_p$	Equivalence class in $\mathcal{G}$ defined for a prototype p
$\mathcal{G}/T(\mathcal{T})$	Set of equivalent classes of $\mathcal{G}$ defined by $T(\mathcal{T})$

Symbol	Description

$\tilde{f}(g)$ — Invariant feature
$\chi(t)$ — Weight or modulus of a relative invariant
$\tilde{\mathcal{F}}(g)$ — Complete system of invariants

Section 3.2: Application-oriented definitions

Symbol	Description
$g(\mathbf{x})$	Continuous image
$\mathbf{x}$	$\mathbf{x} = (x,y)^{\mathrm{T}}$ continuous 2D coordinates: $\mathbf{x} \in \mathcal{X}^2$
$M' \times N'$	Size of a continuous image ($0 \le x < M'$ and $0 \le y < N'$)
$\mathcal{K}$	Set of all possible acquisition parameters
$\mathbf{k}$	Acquisition parameter vector $\mathbf{k} \in \mathcal{K}$
$g(\mathbf{x},\mathbf{k})$	Continuous image
$g_{mn}(\mathbf{k})$	Discrete image
(m,n)	Discrete 2D coordinates $(m,n) := (m\Delta x, n\Delta y)$
$M \times N$	Size of a discrete image ($M = M'/\Delta x$, $N = N'/\Delta y$)
$\mathrm{R}(\varphi)$	2D rotation transformation
φ	Continuous rotation angle $\varphi \in [0, 2\pi)$
$\mathcal{A}$	Set of rotation angles $\varphi \in \mathcal{A}$
$\mathrm{R}(\mathcal{A})$	Set of 2D rotation transformations
$\mathrm{M}(\tau_x, \tau_y, \varphi)$	2D Euclidean motion transformation
$\mathcal{M}$	Set of transformation parameter vectors $(\tau_x, \tau_y, \varphi)$
τ_x, τ_y	Continuous translation parameters in x- and y-direction
$\mathrm{M}(\mathcal{M})$	Set of 2D Euclidean motion transformations
$i\Delta x, j\Delta y$	Discrete translation parameters in x- and y-direction
$k\Delta\varphi$	Discrete rotation angle $k_\varphi := k\Delta\varphi$
K	$K = 2\pi/\Delta\varphi$
M_{ijk}	Discrete cyclical 2D Euclidean transformation
R_k	Discrete cyclical 2D rotation transformation

Chapter 4: Image acquisition under variable illumination direction

Symbol	Description
$\mathcal{B}$	Illumination space
$\mathbf{b}$	Illumination direction $b = (\omega, \theta)^{\mathrm{T}}$

Symbol	Description
ω	Illumination azimuth
θ	Illumination elevation angle
$\mathbf{n}$	Local normal vector of the surface
ϕ	Light incidence angle with respect to $\mathbf{n}$
ϕ'	Light reflection angle with respect to $\mathbf{n}$
$\mathcal{B}_{\theta'}$	Illumination subspace for all $\theta = \theta'$
$\mathcal{B}'$	Illumination subspace: sampling of $\mathcal{B}_{\theta'}$
$\mathcal{S}_g$	Series of images
b	Index of the images in the series $\omega = b\Delta\omega$
B	Number of images in a series $B = 2\pi/\Delta\omega$
g_{mnb}	Discrete image of a series of images

Chapter 5: Extracting invariant features

Section 5.1: Invariant features through integration

Symbol	Description
$\mathcal{U}_g$	Neighborhood of g; $\mathcal{U}_g \subset \mathcal{G}$
$\mathcal{O}_p$	Orbit of patterns equivalent to p
$C_o(\mathcal{G})$	Compact support of $\mathcal{G}$
F	Functional $F : C_0(\mathcal{G}) \longrightarrow \mathbb{R}$
$Z(t)$	Induced transformation on $C_o(\mathcal{G})$
μ	Haar measure
$f(g, \mathbf{w}^l)$	Kernel function of the Haar integral
$\mathbf{w}^l$	Parameter vector of the kernel function
$\mathcal{W}$	Set of parameter vectors $\mathbf{w}_l \in \mathcal{W}$
L	Dimension of $\tilde{\mathcal{F}}(g)$

Section 5.2: Extending the Haar integral for series of images

Symbol	Description
$\mathcal{S}'_g$	Transformed discrete series of images $\mathcal{S}'_g = M_{ijk}\mathcal{S}_g$
$(m', n')^{\mathrm{T}}$	Transformed coordinates of an image $M_{ijk}\, g_{mnb} = g_{m'n'b}$
g'_{mnb}	Transformed image of a series $g'_{mnb} = g'_{m'n'b'}$
$\tilde{\mathbf{f}}^l(\mathcal{S}_g)$	Vectorial kernel function
$f_l^q(\mathcal{S}_g)$	q-th element of the kernel function $\tilde{\mathbf{f}}^l(\mathcal{S}_g)$

Symbol	Description
Q_l	Number of elements of $\mathbf{f}^l(\mathcal{S}_g)$
$r_{1,l}$	Element of $\mathbf{w}^l$
$r_{2,l}$	Element of $\mathbf{w}^l$
α_l	Element of $\mathbf{w}^l$
β_l	Element of $\mathbf{w}^l$
a_l	Element of $\mathbf{w}^l$
$\Delta\rho_l$	Element of $\mathbf{w}^l$
$\mathbf{u}_l^q$	Vector of coordinates to calculate the kernel function
$\mathbf{v}_l^q$	Vector of coordinates to calculate the kernel function
$\mathbf{f}_{ijk}^l$	Transformed vectorial kernel function
f_{lijk}^q	q-th transformed element of the kernel function
$\tilde{\mathbf{f}}^l(\mathcal{S}_g)$	Vectorial invariant feature
$\tilde{f}_l^q(\mathcal{S}_g)$	q-th element of the vectorial invariant
	Section 5.3: Construction of invariant features
$\tilde{\mathbf{f}}_{ij}^l(\mathcal{S}_g)$	Rotation invariant vector
$\tilde{f}_{lij}^q(\mathcal{S}_g)$	q-th element of the rotation invariant vector
$\mathbf{H}_l^q(\mathcal{S}_g)$	Matrix of size $M \times N$, whose elements are given by $\tilde{f}_{lij}^q(\mathcal{S}_g)$
$\mathcal{H}^l$	Sequence of matrices $\mathbf{H}_l^q$
$\mathcal{H}_{ld}^q$	Fuzzy histogram from $\mathbf{H}_l^q$
D	Number of histogram bins
$\hat{f}$	Maximal possible value of $\tilde{f}_{lij}^q(\mathcal{S}_g)$
Δ	Range of the histogram bins
z^d	Exponential function of the weighted histograms $z \in \mathbb{N}$
L_1	Manhattan distance
$\mathcal{S}_{g(k\Delta\varphi)}$	Series of images rotated on $k\Delta\varphi$: $\mathcal{S}_{g(k\Delta\varphi)} := \mathrm{M}_{00k}\mathcal{S}_g$
$\tilde{\mathbf{f}}_{k\Delta\varphi}^l$	Invariant resulting from $\mathcal{S}_{g(k\Delta\varphi)}$
$\mathcal{S}_{g(ij)}$	Series of images translated on i and j: $\mathcal{S}_{g(ij)} := \mathrm{M}_{ij0}\mathcal{S}_g$
$\tilde{\mathbf{f}}_{ij}^l$	Invariant resulting from $\mathcal{S}_{g(ij)}$
e	Linear translation index

Symbol	**Description**

Chapter 6: Classification

Section 6.1: Training process

$\tilde{\mathcal{F}}_g$	System of invariants from $\mathcal{S}_g$: $\tilde{\mathcal{F}}_g = \tilde{\mathcal{F}}(\mathcal{S}_g)$
$\mathcal{W}_g$	Windows of size $M'' \times N'' \times B$; $\mathcal{W}_g \subset \mathcal{S}_g$
ϑ	Overlap factor for the extraction of $\mathcal{W}_g$
W	Number of extracted windows
$\tilde{\mathcal{F}}_{gw}$	System of invariants from a window $\mathcal{W}_g^w$
$\mathcal{L}$	Representative set of series of images
S	Number of series of images in $\mathcal{L}$
c_{gw}	Class corresponding to the window $\mathcal{W}_g^w$
$\breve{g}_{mn}$	Mask of $\mathcal{S}_g$
$\breve{m}_g^w$	Windows extracted from the mask $\breve{g}_{mn}$

Section 6.2: Testing process

$\hat{c}_{gw}$	Predicted class for a window $\mathcal{W}_g$
CM	Confusion matrix
tn	True negative rate
fn	False negative rate
tp	True positive rate
fp	False positive rate
tc	True classification rate
fd	True detection rate

Bibliography

[BB00]   BENNETT, KRISTIN and ERIN BREDENSTEINER: *Duality and geometry in SVM classifiers*. In LANGLEY, P. (editor): *Proceedings of the Seventeenth International Conference on Machine Learning*, pages 57–64, San Francisco, CA, USA, 2000. Morgan Kaufmann Publishers Inc.

[BBL02]   BODNAROVA, ADRIANA, M. BENNAMOUN and S. LATHAM: *Optimal Gabor filters for textile flaw detection*. Pattern Recognition, 35:2973–2991, 2002.

[BKP01]   BEYERER, JÜRGEN, DORIS KRAHE and FERNANDO PUENTE LEÓN: *Characterization of cylinder bores*. In MAINSAH, EVARISTUS, JEFFREY WOOD and DEREK G CHETWYND (editors): *Metrology and Properties of Engineering Surfaces*, chapter 10. Kluwer Academic Publishers, Boston, 2001.

[BP05]   BEYERER, JÜRGEN and FERNANDO PUENTE LEÓN: *Bildoptimierung durch kontrolliertes aktives Sehen und Bildfusion*. Automatisierungstechnik, 53(10):493–502, 2005.

[BP07]   BARSKY, SVETLANA and MARIA PETROU: *Surface texture using photometric stereo data: classification and direction of illumination detection*. Journal of Mathematical Imaging and Vision, 29:185–204, 2007.

[BS01]   BURKHARDT, H. and S. SIGGELKOW: *Invariant features in pattern recognition - fundamentals and applications*. In KOTROPOULOS, C. and I. PITAS (editors): *Nonlinear model-based image/video processing and analysis*. John Wiley and Sons Inc., New York, 2001.

[BSPC05]   BROADHURST, ROBERT E., JOSHUA STOUGH, STEPHEN M. PIZER and EDWARD L. CHANEY: *Histogram statistics of *local model-relative image regions*. In *International Workshop on Deep Structure, Singularities and Computer Vision*, pages 72–83, Maastricht, The Netherlands, June 2005.

[Bur98]   BURGES, CHRISTOPHER: *A tutorial on Support Vector Machines for pattern recognition*. Data Mining and Knowledge Discovery, 2:121–167, 1998.

[BWBK97]   BODNAROVA, ADRIANA, JOHN A. WILLIAMS, MOHAMMED BENNAMOUTI and KURT K. KUBIK: *Optimal textural features for flaw detection on textile materials*. In *In Proceedings of IEEE TENCON'97 Conference*, volume 1, pages 307–310, Brisbane, Australia, December 1997.

[Car40]   CARTAN, HENRI: *Sur la measure de Haar*. C. R. Acad. Sci. Paris, 211:759–762, 1940.

[CFA91]   COHEN, FERNAND S., ZHIGANG FAN and STEPHANE ATTALI: *Automated inspection of textile fabrics using textural models*. IEEE Transactions on Pattern Analysis and Machine Intelligence, 13(8):803–809, 1991.

[Cha94]   CHANTLER, MICHAEL J.: *The effect of variation in illuminant direction on texture classification*. PhD thesis, Heriot-Watt University, August 1994.

[Cha95]   CHANTLER, MICHAEL J.: *Why illumination direction is fundamental to texture analysis*. IEE Proc.-Vis. Image Signal Process, 142(4):199–206, 1995.

[CJ88]   CHEN, JIAHAN and ANIL K. JAIN: *A structural approach to identify defects in textured images*. In *Proceedings of the 1988 IEEE International Conference on Systems, Man, and Cybernetics*, volume 1, pages 29–32, Beijing and Shenyang, China, August 1988.

[CL01]   CHANG, CHIH-CHUNG and CHIH-JEN LIN: *LIBSVM: a library for Support Vector Machines*. Software available at http://www.csie.ntu.edu.tw/~cjlin/libsvm. National Taiwan University, 2001.

[CLIM05]   CHARRON, CYRIL, OUIDDAD LABBANI-IGBIDA and
EL MUSTAPHA MOUADDIB: *Qualitative localization using
omnidirectional images and invariant features*. In *IEEE/RSJ
International Conference on Intelligent Robots and Systems*,
pages 3009–3014, Alberta, Canada, August 2005.

[CM00]   CHANTLER, MICHAEL J. and G. MCGUNNIGLE: *The
response of texture features to illuminant rotation*. In
*Proceedings of the 15th International Conference on Pattern
Recognition*, volume 3, pages 943–946, Barcelona, Spain,
September 2000.

[CP00]   CHAN, CHI-HO and GRANTHAM K. H. PANG: *Fabric defect
detection by Fourier analysis*. IEEE Transactions on Industry
Appllications, 36(5):1267–1276, 2000.

[CP02]   CONCI, AURA and CLAUDIA BELMIRO PROENÇA: *A
system for real-time fabric inspection and industrial decision*. In
*Proceedings of the 14th International Conference on Software
Engineering and Knowledge Engineering*, volume 27, pages
707–714, Ischia, Italy, 2002. ACM.

[CRL94]   CHANTLER, MICHAEL J., G. T. RUSSEL and L. M.
LINNETT: *Illumination: a directional filter of texture*. In
HANCOCK, EDWIN (editor): *Proceedings of the British
Machine Vision Conference BMVC94*, pages 448–458, New
York, September 1994.

[CSPM02]   CHANTLER, MICHAEL J., M. SCHMIDT, M. PETROU and
G. MCGUNNIGLE: *The effect of illuminant rotation on texture
filters: Lissajous's ellipses*. In *Proceedings of the 7th European
Conference on Computer Vision-Part III*, volume 2352, pages
289–303, London, UK, 2002. Springer-Verlag.

[CW00]   CHANTLER, MICHAEL J. and J. WU: *Rotation invariant
classification of 3D surface textures using photometric stereo
and surface magnitude spectra*. In *Proceedings of the British
Machine Vision Conference BMVC2000*, volume 2, pages
486–495, Bristol, UK, September 2000.

[DC70]   DIEUDONNÉ, JEAN and JAMES CARRELL: *Invariant theory old and new.* Academic Press, New York, 1970.

[DC05]   DRBOHLAV, ONDREJ and MICHAEL J. CHANTLER: *Illumination-invariant texture classification using single training images.* In *Texture 2005: Proceedings of the 4th International Workshop on Texture Analysis and Synthesis,* pages 31–36, Beijing, China, October 2005.

[DHS01]   DUDA, RICHARD, PETER HART and DAVID STORK: *Pattern classification.* Wiley-Interscience Publication, New York, 2nd edition, 2001.

[DKH$^+$92] DAMBERK, PAUL, KEVIN KELLY, JOSHUA HELTZER, MARIA L'ANNUNZIATA and THOMAS SMITH: *A guide to pollution prevention for wood furniture finishing.* Technical Report, TUFTS University. Department of Civil Engineering Medford, Massachusetts, 1992.

[Dun73]   DUNN, J. C.: *Continuous group averaging and pattern classification problems.* SIAM Journal on Computing, 2(4):253–271, December 1973.

[EHE05]   ELBEHIERY, H., A. HEFNAWY and M. ELEWA: *Surface defects detection for ceramic tiles using image processing and morphological techniques.* In *Proceedings of World Academy of Science, Engineering and Technology,* volume 5, pages 158–162, Madrid, Spain, May 2005.

[FB06]   FEHR, JANIS and HANS BURKHARDT: *Phase based 3D texture features.* Pattern Recognition, 4174:263–272, 2006.

[FKK07]   FENG, S., H. KRIM and I. A. KOGAN: *3D face recognition using Euclidean integral invariants signature.* In *IEEE/SP 14th Workshop on Statistical Signal Processing,* pages 156–160, Madison, Wisconsin, USA, August 2007.

[Fle77]   FLEMING, WENDELL: *Functions of several variables.* Springer, New York, 2nd edition, 1977.

[FP03]   FORSYTH, DAVID A. and JEAN PONCE: *Computer vision. A modern approach.* Pearson Education International, New Jersey, 2003.

[FRKB05]   FEHR, JANIS, OLAF RONNEBERGER, HAYMO KURZ and HANS BURKHARDT: *Self-learning segmentation and classification of cell-nuclei in 3D volumetric data using voxel-wise gray scale invariants.* Pattern Recognition, 3663:377–384, 2005.

[GBC01]   GARCIA, R., J. BATTLE and X. CUFI: *A system to evaluate the accuracy of a visual mosaicking methodology.* In *MTS/IEEE Conference and Exibition OCEANS 2001*, volume 4, pages 2570–2583, Honolulu, Hawaii, 2001.

[GS02]   GOLDSCHMIDT, ARTUR and HANS-JOACHIM STREITBERGER: *BASF-Handbuch Lackiertechnik.* BASF, Hannover, 2002.

[GSCS08]   GUAN, SHENGQI, XIUHUA SHI, HAIYING CUI and YUQIN SONG: *Fabric defect detection on wavelet characteristics.* In ZHANG, YANDUO, HONGHUA TAN and QI LUO (editors): *IEEE Pacific-Asia Workshop on Computational Intelligence and Industrial Application*, volume 1, pages 366–370, Wuhan, China, December 2008.

[Gun98]   GUNN, STEVE: *Support Vector Machines for classification and regression.* Technical Report, University of Southampton, 1998.

[GW02]   GONZALEZ, RAFAEL and RICHARD WOODS: *Digital image processing.* Prentice Hall, New Jersey, 3rd edition, 2002.

[Haa33]   HAAR, ADOLF: *Der Massbegriff in der Theorie der kontinuierlichen Gruppen.* Mathematische Annalen, 2(34):147–169, 1933.

[Har79]   HARALICK, ROBERT: *Statistical and structural approaches to texture.* Proceedings of IEEE, 67(5):786–804, 1979.

[HB04]   HALAWANI, ALAA and HANS BURKHARDT: *Image retrieval by local evaluation of nonlinear kernel functions around salient points*. In KITTLER, JOSEF, MARIA PETROU and MARK NIXON (editors): *Proceedings of the 17th Intenational Conference on Pattern Recognition*, volume 2, pages 955–960, Cambridge, UK, August 2004.

[HB05]   HEIZMANN, MICHAEL and JÜRGEN BEYERER: *Sampling the parameter domain of image series*. In *Image Processing: Algorithms and Systems IV*, volume 5672, pages 23–33, San José, CA, USA, January 2005.

[HC04]   HUANG, YONG and KAP LUK CHAN: *Texture decomposition by harmonics extraction from higher order statistics*. IEEE Transactions on Image Processing, 13(1):1–14, 2004.

[HCL08]   HSU, CHIH-WEI, CHIH-CHUNG CHANG and CHIH-JEN LIN: *A practical guide to Support Vector classication*. http://www.csie.ntu.edu.tw/~cjlin, 2008.

[HHB04]   HAASDONK, BERNARD, ALAA HALAWANI and HANS BURKHARDT: *Adjustable invariant features by partial Haar-integration*. In KITTLER, JOSEF, MARIA PETROU and MARK NIXON (editors): *Proceedings on the 17th International Conference on Pattern Recognition (ICPR2004)*, volume 2, pages 769–774, Cambridge, UK, August 2004.

[HLM06]   HO, YUN-XIAN, MICHAEL LANDY and LAURENCE MALONEY: *How direction of illumination affects visually percceived surface roughness*. Journal of Vision, 6:634–348, 2006.

[Hoa00]   HOADLEY, R. BRUCE: *Understanding wood: A craftsman's guide to wood technology*. Taunton Press, Newtown, CT, USA, 2000.

[HSD73]   HARALICK, ROBERT, K. SHANMUGAN and ITS'HAK DINSTEIN: *Textural features for image classification*. IEEE Transactions on Systems, Man, and Cibernetics, 3(6):610–621, 1973.

[Hur97] HURWITZT, A.: *Über die Erzeugung der Invarianten durch Integration.* Nachr. Akad. Wiss. Göttingen, pages 71–89, 1897.

[HVB05] HAASDONK, B., A. VOSSEN and H. BURKHARDT: *Invariance in kernel methods by Haar-integration Kernels.* Image Analysis, 3540, 2005.

[Iiv00] IIVARINEN, JUKKA: *Surface defect detection with histogram-based texture features.* SPIE Intelligent Robots and Computer Vision XIX: Algorithms, Techniques, and Active Vision, 4197:140–145, 2000.

[IRV96] IIVARINEN, JUKKA, JUHANI RAUHAMAA and ARI VISA: *Unsupervised segmentation of surface defects.* In *Proceedings of the 13th International Conference on Pattern Recognition,* volume 4, pages 356–360, Viena, Austria, August 1996.

[JMSC04] JIAN, HONGBIN, YI LU MURPHEY, JIANJUN SHI and TZYY-SHUH CHANG: *An intelligent real-time vision system for surface defect detection.* In KITTLER, JOSEF, MARIA PETROU and MARK NIXON (editors): *Proceedings on the 17th International Conference on Pattern Recognition (ICPR2004),* volume 3, pages 239–242, Cambridge, UK, August 2004.

[Kar05] KARRAS, D. A.: *Improved defect detection in textile visual inspection using wavelet analysis and Support Vector Machines.* ICGST International Journal on Graphics, Vision and Image Processing, 5(4):41–47, 2005.

[KMM$^+$94] KITTLER, J., R. MARIK, M. MIRMEHDI, M.PETROU and J. SONG: *Detection of defects in colour texture surfaces.* In *Proceedings of Machine Vision Applications,* pages 558–567, Kawasaki, Kanagawa, Japan, December 1994.

[KP00] KUMAR, AJAY and GRANTHAM K. H. PANG: *Fabric defect segmentation using multichannel blob detector.* Optical Engineering, 39(12):3176–3190, 2000.

[KP02a]   KUMAR, AJAY and GRANTHAM K. H. PANG: *Defect detection in textured materials using Gabor filters*. IEEE Transactions on Industry Applications, 38(2):425–440, 2002.

[KP02b]   KUMAR, AJAY and GRANTHAM K. H. PANG: *Defet detection in textured materials using optimized filters*. IEEE Transactions on Systems, Man, and Cybernetics, 32(5):553–570, 2002.

[Kum03]   KUMAR, AJAY: *Neuronal network based detection of local textile defects*. Pattern Recognition, 36:1645–1659, 2003.

[Lan91]   LANDY, MICHAEL: *Computational models of visual processing*. MIT Press, Cambridge, 1991.

[LB03]   LLADÓ BARDERA, XAVIER: *Texture recognition under varying imaging geometries*. PhD thesis, Universitat de Girona, October 2003.

[LBH05]   LIN, WEI-YANG, NIGEL BOSTON and YU HEN HU: *Summation invariant and its applications to shape recognition*. In *Proceedings of the IEEE International Conference on Acoustics, Speech, and Signal*, volume 5, pages 205–208, Philadelphia, Pennsylvania, USA, March 2005.

[Len90]   LENZ, REINER: *Group theoretical methods in image processing*. Lecture Notes in Computer Science. Springer, Berlin, 1990.

[LICM06]   LABBANI-IGBIDA, OUIDDAD, CYRIL CHARRON and EL MUSTAPHA MOUADDIB: *Extraction of Haar integral features on omnidirectional images: application to local and global localization*. In FRANKE, KATRIN, KLAUS-ROBERT MÜLLER, BERTRAM NICKOLAY and RALF SCHÄFE (editors): *Proceedings of the 28th DAGM-Symposium*, volume 4174, pages 334–343, Berlin, Germany, September 2006.

[Lin09]   LINDNER, CHRISTOPH: *Segmentierung von Oberflächen mittels variabler Beleuchtung*. PhD thesis, Technische Universität München. Shaker Verlag, Aachen, 2009.

[LM06]   LIU, J. JAY and JOHN F. MACGREGOR: *Estimation and monitoring of product aesthetics: application to manufacturinf of "engineered stone" countertops.* Machine Vision and Applications, 16(6):374–383, 2006.

[LP06]   LINDNER, CHRISTOPH and FERNANDO PUENTE LEÓN: *Segmentierung strukturierter Oberflächen mittels variabler Beleuchtung.* Technisches Messen, 73(4):200–207, 2006.

[LSP07]   LINDNER, CHRISTOPH, FABIEN SCHAEFFLER and FERNANDO PUENTE LEÓN: *Texture-based surface segmentation using second-order statistics of illumination series.* In *Informatik 2007: Informatik trifft Logistik–Beiträge der 37. Jahrestagung der Gesellschaft für Informatik*, volume 109 of *GI Proceedings*, pages 32–37, Bremen, 2007.

[LVP05]   LOPEZ, F., J. VALIENTE and J. PRATS: *Surface grading using soft colour-texture descriptors.* In *Proceedings of the 10th Iberoamerican Congress on Pattern Recognition*, pages 13–23, Havana, Cuba, November 2005.

[LWY97]   LIN, HSI-CHIH, LING-LING WANG and SHI-NINE YANG: *Extracting periodicity of a regular texture based on autocorrelation functions.* Pattern Recognition Letters, 18:433–443, 1997.

[Mäe03]   MÄENPÄÄ, TOPI: *The local binary pattern approach to texture analysis - extensions and applications.* PhD thesis, Faculty of Technology, University of Oulu. Oulu University Press, Oulu, Finland, August 2003.

[MBR04]   MURINO, VITTORIO, MANUELE BICEGO and IVAN A. ROSSI: *Statistical classification of raw textile defects.* In *Proceedings of the 17th International Conference on Pattern Recognition*, volume 4, pages 311–314, Cambridge, UK, August 2004. IEEE Computer Society.

[MC00]   MCGUNNIGLE, G. and MICHAEL J. CHANTLER: *Rough surface classification using point statistics from photometric stereo.* Pattern Recognition Letters, 21:593–604, 200.

[McG98]   MCGUNNIGLE, G.: *The classification of textured surfaces under varying illuminant direction.* PhD thesis, Heriot-Watt University, June 1998.

[MCH+06]   MANAY, SIDDHARTH, DANIEL CREMERS, BYUNG-WOO HONG, ANTHONY YEZZY and STEFANO SOATTO: *Integral invariants for shape matching.* IEEE Transactions on Pattern Analysis and Machine Intelligence, 28(10):1602–1618, October 2006.

[MF00]   MARSHALL, AMY and JENNIFER FIELDAS: *Case studies: Low-VOC/HAP wood furniture coating.* Technical Report, Midwest Research Institute, Springfield, Virginia, May 2000.

[MGD00]   MALLIK-GOSWAMI, BITHIKA and ASIT K. DATTA: *Detecting defects in fabric with laser-based morphological image processing.* Textile Research Journal, 70:758–762, 2000.

[MMT04]   MONADJEMI, AMIRHASSAN, M. MIRMEHDI and B. THOMAS: *Restructured eigenfilter matching for novelty detection in random textures.* In *Proceedings of the 15th British Machine Vision Conference*, pages 637–646, United Kindom, September 2004.

[MNA+04]   MANDRIOTA, C., N. NITTI, N. ANCONA, E. STELLA and A. DISTANTE: *Filter-based feature selection for rail defect detection.* Machine Vision and Applications, 15:179–185, 2004.

[Mon04]   MONADJEMI, AMIRHASSAN: *Towards efficient texture classification and abnormality detection.* PhD thesis, University of Bristol, UK, 2004.

[MP05]   MÄENPÄÄ, TOPI and MATTI PIETIKÄINEN: *Texture analysis with local binary patterns.* In CHEN, C., L. PAU and P. WANG (editors): *Handbook of pattern recognition and computer vision*, pages 197–216. World Scientific, Signapore, 3rd edition, 2005.

[MTP03] MÄENPÄÄ, TOPI, MARKUS TURTINEN and MATTI PIETIKÄINEN: *Real-time surface inspection by texture.* Real-Time Imaging, 9(5):289–296, October 2003.

[NKS02] NISKANEN, MATTI, HANNU KAUPPINEN and OLLI SILVÉN: *Real-time aspects of SOM-based visual surface inspection.* In HUNT, MARTIN A. (editor): *Machine Vision Applications in Industrial Inspection*, volume 4664, pages 123–134, San José, CA, USA, January 2002.

[NSK01] NISKANEN, MATTI, OLLI SILVÉN and HANNU KAUPPINEN: *Color and texture based wood inspection with non-supervised clustering.* In *Proceedings of the 12th Scandinavian Conference on Image Analysis*, pages 336–342, Bergen, Norway, June 2001.

[OD92] OHANIAN, PHILIPPE P. and RICHARD C. DUBES: *Performance evaluation for four classes of textural features.* Pattern Recognition, 25(8):819–833, 1992.

[Oja96] OJALA, TIMO: *Multichannel approach to texture description with feature distributions.* Technical Report, University of Maryland. Center for Automation Research, 1996.

[OPH96] OJALA, TIMO, MATTI PIETIKÄINEN and DAVID HARWOOD: *A compartive study of texture measures with classifcation based on feature distributions.* Pattern Recognition, 29(1):51–59, 1996.

[OPM02] OJALA, TIMO, MATTI PIETIKÄINEN and TOPI MÄENPÄÄ: *Multiresolution gray-scale and rotation invariant texture classification with local binary patterns.* IEEE Transactions on Pattern Analysis and Machine Intelligence, 24(7):971–987, 2002.

[OVOP01] OJALA, TIMO, KIMMO VALKEALAHTI, ERIKKI OJA and MATTI PIETIKÄINEN: *Texture discrimination with multidimensional distributions of signed gray-level differences.* Pattern Recognition, 34:727–739, 2001.

[PAP08]   PÉREZ GRASSI, ANA, MIGUEL ÁNGEL ABIÁN PÉREZ and
           FERNANDO PUENTE LEÓN: *Illumination and model-based
           detection of finishing defects*. In LEÓN, FERNANDO PUENTE
           (editor): *Reports on Distributed Measurement Systems*, pages
           31–51. Shaker Verlag, Aachen, 2008.

[PB05]   PUENTE LEÓN, FERNANDO and JÜRGEN BEYERER:
           Oberflächencharakterisierung durch morphologische Filterung.
           Technisches Messen, 12:663–670, 2005.

[PCP02]   PENIRSCHKE, ANDREAS, MICHAEL J. CHANTLER and
           MARIA PETROU: *Illuminant rotation invariant classification of
           3D surface textures using Lissajous's ellipses*. In *2nd
           International Workshop on Texture Analysis and Synthesis*,
           pages 103–107, Copenhagen, June 2002.

[Per03]   PERNKOPF, FRANZ: *Detection of surface defects on raw steel
           blocks using Bayesian network classifiers*. Pattern Analysis
           and Applications, 7:333–342, 2003.

[POX00]   PIETIKÄINEN, M., T. OJALA and Z. XU: *Rotation-invariant
           texture classification using feature distributions*. Pattern
           Recognition, 33:43–52, 2000.

[PPMC06]   PISTORI, HEMERSON, WILLIAM A. PARAGUASSU,
           PRISCILA S. MARTINS and MAURO P. CONTI: *Defect
           detection in raw hide and wet blue leather*. In *CompIMAGE -
           Computational Modelling of Objects Represented in Images:
           Fundamentals, Methods and Applications*, Coimbra, Portugal,
           October 2006.

[PR03]   PUENTE LEÓN, FERNANDO and NORBERT RAU: *Detection
           of machine lead in ground sealing surfaces*. Annals of the
           CIRP, 52(1):459–462, 2003.

[PTP+07]   PÉREZ GRASSI, ANA, MICHAEL THUY, FERNANDO
           PUENTE LEÓN, HICHEM MAAZOUN and SABRINA
           SCHMIDT: *Klassifikation menschlicher Verkehrsteilnehmer in
           MIR-Bildern: eine erste Annäherung*. In PUENTE LEÓN,
           FERNANDO and MICHAEL HEIZMANN (editors):

Bildverarbeitung in der Mess- und Automatisierungstechnik,
volume 1981, pages 293–302, Düsseldorf, 2007. VDI-Verlag.

[Pue99]   PUENTE LEÓN, FERNANDO: *Automatische Identifikation von
Schußwaffen.* PhD thesis, Fakultät für Maschinenbau der
Universität Karlsruhe (TH). Vol. 787 of Fortschritt-Berichte
VDI, Reihe 8, VDI Verlag, Düsseldorf, 1999.

[Pue02a]   PUENTE LEÓN, FERNANDO: *Evaluation of honed cylinder
bores.* Annals of the CIRP, 51(1):503–506, 2002.

[Pue02b]   PUENTE LEÓN, FERNANDO: *Komplementäre Bildfusion zur
Inspektion technischer Oberflächen.* Technisches Messen,
69(4):161–168, 2002.

[Pue06]   PUENTE LEÓN, FERNANDO: *Automated comparison of
firearm bullets.* Forensic Science International, 156(1):40–50,
2006.

[Ral02]   RALLÓ, MIQUEL: *Wavelet based techniques for textile
inspection.* In *Workshop on Wavelets and Applications*, pages
433–466, Barcelona, Spain, July 2002.

[RB06]   REISERT, M. and H. BURKHARDT: *Invariant features for
3D-data based on group integration using directional
information and spherical harmonic expansion.* In *Proceedings
of the 18th International Conference on Pattern Recognition
(ICPR'06)*, volume 4, pages 206–209, Hong Kong, August
2006. IEEE Computer Society.

[RBS02]   RONNEBERGER, OLAF, HANS BURKHARDT and ECKART
SCHULTZ: *General-purpose object recognition in 3D volume
data sets using gray-scale invariants – Classification of airborne
pollen-grains recorded with a confocal laser scanning
microscope.* In *Proceedings of the 16th International Conference
on Pattern Recognition*, volume 2, pages 209–295, Quebec,
Canada, August 2002.

[RFB05]   RONNEBERGER, OLAF, JANIS FEHR and HANS
BURKHARDT: *Voxel-wise gray scale invariants for*

simultaneous segmentation and classification. Pattern Recognition, 3663:85–92, 2005.

[SB98] SIGGELKOW, S. and H. BURKHARDT: *Image retrieval based on local invariant features*. In *Processings of the IASTED International Conference on Signal and Image*, Las Vegas, Nevada, USA, October 1998.

[Sch01] SCHAEL, M.: *Texture defect detection using invariant textural features*. In *Proceedings of the 23rd DAGM-Symposium on Pattern Recognition*, volume 2191, pages 17–24, London, UK, 2001. Springer-Verlag.

[Sch05] SCHAEL, M.: *Methoden zur Konstruktion invarianter Merkmale für die Texturanalyse*. PhD thesis, Albert-Ludwigs-Universität Freiburg, 2005.

[SHW88] SIEW, LEE HOK, ROBERT M. HODGSON and ERROL J. WOOD: *Texture measures for carpet wear assessment*. IEEE Transactions on Pattern Analysis and Machine Intelligence, 10:92–105, 1988.

[SIB05] SETIA, LOKESH, JULIA ICK and HANS BURKHARDT: *SVM-based relevance feedback in image retrieval using invariant feature histograms*. In *Proceedings of the IAPR Conference on Machine Vision Applications*, pages 542–545, Tsukuba Science City, Japan, May 2005.

[Sig02] SIGGELKOW, S.: *Feature histogram for content-based image retrieval*. PhD thesis, Albert-Ludwigs-Universität Freiburg, 2002.

[SM95a] SCHULZ-MIRBACH, HANNS: *Anwendung von Invarianzprinzipien zur Merkmalgewinnung in der Mustererkennung*. PhD thesis, Technische Universität Hamburg-Harburg. Reihe 10, Nr. 372, VDI Verlag, Februaury 1995.

[SM95b] SCHULZ-MIRBACH, HANNS: *Invariant features for gray scale images*. Technical Report, Institut für technische Informatik

I, Technische Universität Hamburg-Harburg, 21071
Hamburg, Germany, 1995.

[SS96] SCHEITHAUSER, MARGOT and HANS-JÜRGEN SIRCH:
Filmfehler an Holzlacken. Vincentz Verlag, Hannover, 1996.

[SS99] SIGGELKOW, S. and M. SCHAEL: *Fast estimation of invariant
features*. In *Mustererkennung, DAGM 1999*, pages 181–188,
London, UK, September 1999. Springer-Verlag.

[SS00] SCHAEL, MARC and SVEN SIGGELKOW: *Invariant
grey-scale features for 3D sensor-data*. In *15th International
Conference on Pattern Recognition*, volume 2, pages 531–535,
Barcelona, Spain, September 2000. IEEE Computer Society.

[SSR$^+$06] SCHULZ, JANINA, THORSTEN SCHMIDT, OLAF
RONNEBERGER, HANS BURKHARDT, TARAS PASTERNAK,
ALEXANDER DOVZHENKO and KLAUS PALME: *Fast scalar
and vectorial grayscale based invariant feature for 3D cell nuclei
localization and classification*. Pattern Recognition,
4174:182–191, 2006.

[SSY79] SCHVEDOV, A., A. SCHMIDT and V. YAKUBOVICH:
Invariant systems of features in pattern recognition.
Automation Remote Control, 40:131–142, 1979.

[TBH01] TSAI, D. M. and B. B. HSIAO: *Automatic surface inspection
using wavelet reconstruction*. Pattern Recognition,
34:1285–1305, 2001.

[TH99] TSAI, D. M. and C. Y. HSIEH: *Automated surface inspection
for directional textures*. Image and Vision Computing,
18(1):49–62, 1999.

[TH03] TSAI, D. M. and Y. HUANG: *Automated surface inspection
for statistical textures*. Image and Vision Computing,
21:307–323, 2003.

[TJ99] TUCERYAN, MIHRAN and ANIL K. JAIN: *Texture analysis*.
In CHEN, C., L. PAU and P. WANG (editors): *The handbook
of pattern recognition and computer vision*, pages 207–248.
World Scientific, Signapore, 2 edition, 1999.

[TK06] THEODORIDIS, SERGIOS and KONSTANTINOS KOUTROUMBAS: *Pattern recognition*. Academic Press, Amsterdam, 2006.

[TL08] TRUCHETET, FRED and OLIVER LALIGANT: *A review on industrial applications of wavelet and multiresolution based signal-image processing*. Journal Electronic Imaging, 17(3):03–11, July 2008.

[TPFP08] THUY, MICHAEL, ANA PÉREZ GRASSI, VADIM A. FROLOV and FERNANDO PUENTE LEÓN: *Fusion von MIR-Bildern und Lidardaten zur Klassifikation menschlicher Verkehrsteilnehmer*. In MAURER, MARKUS and CHRISTOPH STILLER (editors): *5. Workshop Fahrerassistenzsysteme*, pages 168–175, Walting, Altmühltal, 2008. Freundeskreis Mess- und Regelungstechnik Karlsruhe e.V.

[TRB07] TEMERINAC, MAJA, MARCO REISERT and HANS BURKHARDT: *Invariant features for searching in protein fold databases*. International Journal of Computer Mathematic, 84(5):635–651, 2007.

[TW00] TSAI, D. M. and S. K. WU: *Automated surface inspection using gabor filters*. The International Journal of Advanced Manufacturing Technology, 16:474–482, 2000.

[Vap79] VAPNIK, VLADIMIR: *Estimation of dependences based on empirical data (in Russian)*. Nauka, Moskow. English translation 1982 Spriger Verlag, New York, 1979.

[Vil68] VILENKIN, N. J.: *Special functions and the theory of group representation*. American Mathematical Society, Providence, RI, 1968.

[VM05] VIGNOTE PEÑA, SANTIAGO and ISAAC MARTINEZ ROJA: *Tecnología de la madera*. Mundi-Prensa, Madrid, 3rd edition, 2005.

[vN36] NEUMANN, JOHN VON: *The uniquess of Haar's measure*. Mat. Sbornik, 1:721–734, 1936.

[VNP86] VILNROTTER, FELICIA M., RAMAKANT NEVATIA and KEITH E. PRICE: *Structural analysis of natural textures.* IEEE Transactions on Pattern Analysis and Machine Intelligence, 8(1):76–89, 1986.

[WC03] WU, JIAHUA and MICHAEL J. CHANTLER: *Combining gradient and albedo data for rotation invariant classification of 3D surface texture.* In *Proceedings of the 9th IEEE iInternational Conference on Computer Vision*, volume 2, pages 848–855, Nice, France, October 2003. IEEE Computer Society.

[Wei40] WEIL, ANDRÉ: *L'intégration dans les groupes topologiques et ses applications.* Hermann, Paris, 1940.

[Wil99] WILLIAMS, SAM: *Wood hadbook: wood as an engineering material.* USDA Forest Service, Forest Products Laboratory, Madison, Wisconsin, USA, 1999.

[Woo80] WOODHAM, ROBERT J.: *Photometric method for determining surface orientation from multiple images.* Optical Engineering, 1:139–144, January 1980.

[Woo90] WOOD, ERROL J.: *Applying Fourier and associated transforms to pattern characterization in textiles.* Textile Research Journal, 60:212–220, 1990.

[Woo96] WOOD, JEFFREY: *Invariant pattern recognition: a review.* Pattern Recognition, 29(1):1–17, 1996.

[WPL00] WILSCHI, KLAUS, AXEL PINZ and TONY LINDEBERG: *An automatic assessment scheme for steel quality inspection.* Machine Vision Applications, 12:113–128, 2000.

[WX99] WEN, W. and A XIA: *Verifying edges for visual inspection purposes.* Pattern Recognition Letters, 20:315–328, 1999.

[XMT06] XIE, XIANGHUA, MAJID MIRMEHDI and BARRY THOMAS: *Colour tonality inspection using eigenspace features.* Machine Vision and Applications, 16(6):364–373, 2006.

[YPY05]   YANG, X., GRANTHAM K. H. PANG and N. YUNG: *Robust fabric defect detection and classification using multiple adaptive wavelets*. IEE Proceedings-Vision, Image and Signal Processing, 152(6):715–723, 2005.

[YZC⁺08]   YANG, WEI, SU ZHANG, YAZHU CHEN, YAQING CHEN, WENYING LI and HONGTAO LU: *Effective shape measures in malignant risk assessment for breast tumor on sonography*. In *International Multi-symposium on Computer and Computational Sciences*, pages 51–56, Washington, DC, USA, October 2008.

[ZKL08]   ZHAO, FENGDA, LINGFU KONG and XIANSHAN LI: *LSRII feature based particle filter localization for mobile robot*. In *Proceedings of the 7th World Congress on Intelligent Control and Automation*, pages 2350–2354, Chongqing, China, June 2008.